Tiny Dynamo

HOW ONE OF THE WORLD'S SMALLEST COUNTRIES
IS PRODUCING SOME OF OUR MOST IMPORTANT INVENTIONS

* **Marcella Rosen** with David Kornhaber

Copyright © 2012 Untold News

ISBN 978-0-9887973-0-7
Library of Congress Control Number: 2012956248

For book inquiries or bulk orders
please email: *contact@untoldnews.org*

Printed in the USA

Second Printing: January 2013
Third Printing: September 2013

Contents

Acknowledgments

I would like to thank David Kornhaber for his enormous contribution to the manuscript. David's uncanny ability to synthesize complicated subjects so they are easy to understand, coupled with his modern perspective and storytelling flair, helped to bring a vastly improved *Tiny Dynamo* to fruition.

I'd also like to thank Margie Fox for providing the initial inspiration for the book, for contributing crucial advice regarding its design and structure, and for serving as a project-long cheerleader.

Thanks, too, to Megan Flood for her dynamic and original cover design.

Finally, my gratitude to Anne Etra for spearheading the effort to bring the book to its widest possible audience.

Without the contributions of these individuals *Tiny Dynamo* would not have come to pass.

Preface

While everyone has been focused on the country's decades of military conflicts, Israel has quietly become the most energetic, ambitious, go-go incubator of entrepreneurialism and invention the planet has ever seen.

It's true: Israel is a barrier-breaking dynamo of a kind never before witnessed in history. Acre-for-acre, citizen-for-citizen, no place is churning out more ideas, more products, more procedures and devices and technologies than this tiny strip of land along the Mediterranean.

And the work that Israel is turning out is saving and improving lives around the world, every day.

I could give you statistics: Israel is home to more start-ups, inventions and patents than the entirety of the EU; Israel attracts easily twice as much venture capital per capita as the next nearest recipient (the USA).

I could give you more. I could overwhelm you with statistics.

But statistics you can look up. What's more important than the numbers is that Israel's energy and inventiveness and output matter. From the hospital to the farm, from outer space to your kitchen, Israel's life-saving, life-giving, life-enhancing creations make a positive difference every day in your life, in the lives of people you care about and in the lives of people you'll never meet.

It's my firm belief that the world needs to hear this story — about a tiny nation where an unprecedented degree of life-altering work is being done every day. Further, the world shouldn't wait to hear it.

In fact, with all the turmoil taking place in the Middle East, I think the more information people have about what's going on beyond the headlines, the better.

Now . . . Is Israel perfect? Or course not. There's never been a perfect country and there never will be. And perhaps no one is as aware of this as the average Israeli; the political choices and direction of the country are a source of constant internal tension, debate and even violence between its citizens. Like all vibrant democracies, Israel is constantly defining itself to itself. And as with all democracies, that process can often be ugly to watch — both from within and without.

But, professionally, I come from the world of communications and media, and I know deep in my bones that no country is equivalent to the media coverage of that country. Likewise, it's anathema to me to hear a story half-told. And having traveled to Israel a dozen times — and to many other parts of the world as well — I can say with confidence that the common perception about a country rarely squares with the life that's actually lived in that place, and the people who live it. So if it's true that you should judge a person not by what is said about him but by what he does, then it follows that you should do the same for countries.

Unfortunately, the media doesn't really give you a chance to develop an informed opinion about Israel — or many other countries, for that matter — because, in the press, "If it bleeds it leads" — and the Middle East certainly does bleed. So that's all you hear.

But the truth is this: while the world's attention has for decades been focused on one single dimension of Israeli life, something entirely different has been taking place away from the cameras: Israel has quietly become the little country that changed the world — and your life — for the better . . . without you even knowing it.

This book tells 21 stories about Israelis who are emblematic of their nation's determination to make a positive difference, and the work through which they're expressing that determination. I chose these stories because they represent a cross-section of what's going on within Israel's borders . . . but for every story I chose, there are dozens more that could have been included.

The details of the stories differ, but what I hope this collection replicates for you is something like the epiphany I experienced on one of my earliest trips to Israel.

I was there on business, and in the course of my travels around the country I had come into contact with a number of people who were embarking on entrepreneurial ventures of one kind or another — none connected to each other, nor even in related areas of business.

It happened that one such meeting required me to take a road trip with an Israeli colleague. As we motored along the highway, I looked out one window at a vast and barren desert: Jordan. I looked out the opposite window at green, productive farmland that stretched to the horizon: Israel.

At that moment something clicked, like the tumblers of a lock falling into alignment, as I realized that what's going on in Israel — all those Nobel laureates, all those patents, the green fields in the desert — is so much bigger than the sum of its parts. This little country is a place where achievement happens — achievement with global benefits.

And my skin still tingles when I remember the next thought I had:

If tiny, beleaguered Israel can generate these kinds of results under its current circumstances, imagine what would happen — imagine what it could achieve — if it were released from the shackles of warfare. If this little country of fewer than eight million souls could focus the entirety of its energy and resources and resilience on the problems and puzzles facing us all, how much better a place would this world be?

. . .

It is my wish that this book will encourage you to set aside momentarily what you think and know about Israel and look without preconception at the work being done there, and the people of good will and unwavering commitment who are doing it.

I would then ask that you imagine living in a world where, every day, people's lives are saved and bettered because of this work, and because of these people.

Finally, I would point out that this is a world where we — all of us, around the globe — are already living.

Marcella Rosen

1. Reducing Hospital Infections

See if this statistic blows your mind, because it sure blows mine. Every year, 90,000 people die in American hospitals as a result of bacterial infections they contract while they're in the hospital.

90,000.

Death is going to happen in hospitals, but that number isn't due to car accidents, crime, or cancer. It's due to infection caught in the very place where you go to heal.

90,000.

It's a maddening statistic. Let it sink in.

Then realize that, worldwide, the number is one million.

The causes are numerous. Unsterilized instruments. Unwashed hands. Door handles. Airborne bugs being swapped by close-quartered patients.

But the biggest culprits couldn't be more mundane: Pajamas. Sheets. Pillowcases. Gowns. Curtains.

It turns out that textiles are among the most common carriers and conveyers of the bacteria that cause hospital infections, which enter the body through wounds, incisions, and foreign bodies such as catheters or tubing.

Now I realize that we're never going to entirely eliminate hospital infections. But the first time I heard that sheets and pajamas and the like are the biggest infection malefactors, I thought "Come on, surely we can do better than that."

Aharon Gedanken agrees. And he has developed a method for making anti-bacterial fabrics that he believes will drastically reduce the incidence of hospital infections and save thousands of lives worldwide.

"The general view regarding this problem is that the hospital of the future should be as anti-bacterial as humanly possible," says Professor Gedanken. "But I have developed a technology that can move us a long way toward that goal right now."

Prof. Gedanken's technology represents a major step forward toward eliminating a problem that most people aren't even aware of. And it all started with smelly socks.

Prof. Gedanken is a groundbreaking chemist and director of the Kanbar Laboratory for Nanomaterials at the Bar-Ilan University Institute of Nanotechnology and Advanced Materials.

In 2003 Professor Gedanken was involved in a project to make socks that Israeli soldiers could wear for a week or more between washings. Professor Gedanken realized that the offensive odor from unwashed socks comes from the bacteria that thrive in the warm, moist world inside a soldier's boot. Eliminate the bacteria and you eliminate the smell.

Prof. Gedanken developed a revolutionary method of passing ultrasonic waves through a zinc oxide solution, forming microjets that blast anti-bacterial zinc oxide nanoparticles onto — and into — a substrate introduced into the solution.

In short, if you dunk a sock in the solution, it comes out impregnated with an anti-bacterial "coating" that is actually part of the sock's very fabric.

From the sock project, Prof. Gedanken and his colleagues moved on to the challenge of creating an anti-bacterial bandage that led him to naturally consider the problem of hospital textiles.

"We have shown that this is a very efficient coating," says Prof. Gedanken. "Coating is actually a misleading term. This process forces the coating to penetrate whatever you're treating. Metal, polymer, ceramics, paper. We've tested it on glass, and we can show that it actually penetrates the glass."

That penetrative quality is the key to Prof. Gedanken's results with hospital fabrics.

"When we talk to large hospitals and health care companies, they say to us 'Look, Prof. Gedanken, we have twenty companies like you coming to show us coated textiles and they just don't hold up.'"

That's because hospitals are ruthless in the way they subject their pillowcases and pajamas and so on to fabric's ultimate enemy: the washing machine.

Hospitals run washes at temperatures up to 92 degrees Celsius — that's 197 degrees Fahrenheit, or about 15 degrees short of boiling water.

"Other companies produce coated materials," says Prof. Gedanken, "but when you wash them like that, the coating comes off."

Prof. Gedanken's coating doesn't.

Currently Prof. Gendanken and his partners can efficiently produce the kind of materials hospitals would require for smaller items like pillowcases and pajamas. The next challenge is to create a machine that can produce coated materials that will satisfy a busy hospital's need for sheets, curtains, drapes.

Prof. Gedanken's team has partnered with a consortium of seventeen academic and industrial groups from eleven European countries to produce such a machine, because, says Prof. Gedanken, "even though we can demonstrate the effectiveness of our process, and show that we can produce it successfully at the smaller end of the scale, those in a position to put our materials to work for people can't rely on their imaginations. They need to see it with their own eyes." The consortium gives Prof. Gedanken's work a broad-based, international credibility that should carry weight with those who can put his materials to work on behalf of patients worldwide.

And with the problem of fabric-borne infections gaining recognition inside the medical and industrial communities, it seems like it's only a matter of time before anti-bacterial materials become the standard in hospitals throughout the U.S. and, eventually, around the world.

That might not make checking into the hospital any more pleasant. But it will significantly increase the likelihood that you will eventually check out.

2. New Airport Security

We've all got secrets, and in most cases it's in a free society's interest that they stay secret.

But what happens when the secret a person is harboring is an intent and a plan to do harm to people, and to do it publicly and sensationally?

The truth is we live in a world where certain individuals — in settings all over the world — are bent on committing acts of terrorism, and I believe we need to do everything in our power to stop them before they act — while still respecting the liberties that make our society the freest and most open on Earth.

Therein lies the rub. How do we identify and separate out the do-badders without making the rest of us give up so many of our freedoms that we might as well be in jail too?

Answering that one is like trying to solve an ethical jigsaw puzzle with no clear corners, no defined edges, and no picture on the box to act as a guide.

But Ehud Givon thinks he's found a big piece. For Mr. Givon, it all comes down to the fact that, while you may be able to deceive everyone else you can never really fully deceive yourself.

"Let's say you know who took the last apple from the conference table," Mr. Givon explains. "And let's say that this knowledge is not something you want to share with other people, because in this case it's, say, your boss. You can be successful in hiding this knowledge. You can be great at it. You can have the world's most impenetrable poker face. But you cannot refrain from knowing that you know it."

This simple but significant truth was the inspiration behind what Mr. Givon cheerfully describes as a "lunatic idea."

Mr. Givon is the CEO of WeCU Technologies — with headquarters in Ceasarea. He and his team of widely diversified specialists have developed a "terrorist detection system" that screens out people harboring malicious intent by, essentially, reading their thoughts.

It sounds like sci-fi, but there's nothing fictional about it.

Mr. Givon pulled together a team of seven specialists from fields ranging from behavioral psychology to computer engineering to create the next generation of terrorist detection: the WeCU ("we see you") Security System.

WeCU is a catch-all term for a complex array of software and machines that inject specific visual and audio stimuli into public settings, such as an airline or bus terminal. Those who are innocently going about their business will never know that these stimuli are present. But those with wrongdoing in mind will have an involuntarily physiological reaction to the stimuli: raised blood pressure, increased heart rate, quicker breathing and the like.

These reactions are picked up and analyzed in real time by sensors distributed around the venue. The result is a seamless system that picks out the guilty while letting the innocent carry on without any hassle.

But the development of the system itself was anything but seamless.

It all started in response to the terrorist strikes of September 11, 2001, when Mr. Givon, an engineer by training, assembled a group of his friends and started brainstorming a way to advance the ball in terms of combatting terrorism.

"Like so many people, we wanted to do something about this horrible problem that seemed to be getting out of hand," he recalls, "but everything that was obvious, someone was doing something with it."

Then they approached the problem from a different angle.

"We saw people at the airport surrendering their belts and laptops, their shoes and coats, and knew there had to be a better way," says Givon. "We turned to the notion that people with malicious intent would have to behave differently from other people — and if you could create the right environment, they would reveal themselves without even knowing it."

From that kernel of inspiration grew the determination to create a method that would change the way security systems search for terrorists.

But it wasn't long before they discovered that they were entering completely uncharted territory.

"The more we dug into this, the more we found that we were standing on a platform where nobody had stood before. We were sure we'd find most of the answers we were looking for in the existing literature, but it was completely blank," Mr. Givon recalls.

But rather than fold up and go home, Mr. Givon sunk increasing amounts of his time, energy and money into the operation.

"In engineering, you have a goal. You want to lay a piece of pipe. You know what type of pipe to use, what length and so on. But in development you come to the end of the known technology. But you're sure you're on the right track, you've got a good idea. You have to find some new ideas that nobody has had in the past, and that means bringing in the people to do it if you can't do it yourself. You have to be stubborn, and you have to believe in your idea."

That stubbornness paid off.

As Mr. Givon describes it, "All the complex work resulted in the WeCU — a system that can screen people who are trying to hide a very unique piece of information in a way that is very friendly to other people."

Though he can't reveal proprietary details of how his system works, Mr. Givon describes a typical scenario as taking place "in direct parallel with routine activities."

For example, say you're checking in for a flight at an airport kiosk. The screen in front of you, the signs around you, the PA system, the lighting — all come together to create a specific environment.

WeCU injects certain stimuli into that environment — for instance, through words or images on the kiosk screen, or through PA announcements — that will mean something to those who are plotting to do harm, while meaning nothing to those who are not.

"That comes from one of our two main areas of activity, the behavioral sciences," Mr. Givon asserts.

The other underpinning of the WeCU system is engineering — specifically thermal and biometric sensors that can capture physiological reactions and analyze them in real time, returning reports on individuals who appear to be reacting to the stimuli in time for the authorities to separate that person and examine him or her more closely.

But the most important question is: does it work?

"We started by testing the system on security experts, to see if they could locate the stimuli, and they couldn't" says Mr. Givon proudly. "This goes for the written stimuli, the audio, everything. Only after we show it to them do they get it."

Protecting the specifics of his system prevents Mr. Givon from going into further details about his testing process, but he will say that he and his team have tested the system on more than 1800 people and are ready to deploy it.

Mr. Givon says he has strong interest in his system from

US Homeland Security and agencies from a number of other countries.

If things go as he hopes, it won't be long before Mr. Givon's WeCU system puts a significant dent in the plans of would-be terrorists around the world, and lets the rest of us keep our belts and shoes on in the process.

3. Better Electric Cars

I love my bulletin board.

No really. I recently replaced my old brown school-cafeteria style bulletin board with a slick new model, and this upgraded version neatly demonstrates the power of innovation. My new bulletin board looks great, it delivers excellent performance, it's quiet, and as it's fabricated from recycled rubber, it makes good, productive use out of waste material. That's a lot of bang from something that really exists only to give me somewhere to impale my dry cleaning tickets.

But I mean it when I say that my bulletin board perfectly embodies my belief that a lot of the best ideas are succinct, humble, clean and straightforward.

However . . .

I also adhere to a notion that was best articulated by H.L. Mencken when he said "For every complex problem, there is a solution that is clear, simple and wrong."

Or, put another way: when you're facing a daunting, mountainous, intractable behemoth of a challenge, you gotta go big.

I guess problem size is in the eye of the problem-haver, but for my money, there is no bigger burst water heater in the world's basement than the issue of oil.

Oil is hungry black squid with tentacles wrapped around the world's wallets, politics and environment. Oil drives our foreign policy and our domestic lives, dictates market dynamics and sets special interests against one another.

While oil has enabled us to make progress in a lot of areas, it seems like its time is running out, and I think it's safe to say that aside from those who directly work in the industry, no one really likes oil . . . and even the petro-companies themselves are running ads acknowledging that oil has become troublesome to the world (exhibit A: BP changing its company slogan to "beyond petroleum" — and this was before its Gulf of Mexico blowout).

How dependent are we? According to the US Energy Information Administration, the United States alone consumes about 10 million barrels of petroleum products each day in the form of gasoline alone, and we put it into more than 210 million vehicles traveling over 7 billion miles per day.

Getting people off oil is going to take work on multiple fronts, it's going to take time and effort, and it's going to be a monumental struggle on lots of levels, like a combination between trench warfare and a food fight.

But it's a fight we need to have. And we don't just need someone to go first, we need a lot of people, countries, and companies to go first. Fortunately, one of the first to go first is also one of the first to go big.

Better Place is an international electric car company that is attacking the problem of oil dependency from a lot of directions. Its solution in a nutshell: completely rebuild the automotive infrastructure of every country on earth.

The brainchild of Israeli entrepreneur Shai Agassi, Better Place has entered into a partnership with Israel that will make it the first country to set out on a course toward an all-electric car infrastructure.

Electric cars have been around since before cars were fueled by gas; the first cars were powered by batteries. What Better Place has done is change the way we think about electric cars and bring that thinking more in line with the energy sources of the future.

"When you buy a car, you are buying a machine, but more fundamentally, you are buying mobility. You are buying miles," says Michael Granoff, Head of Oil Independence Policies for Better Place.

"You might not think of it this way, but when you buy a car, you enter into an implicit contract with oil companies to provide you with those miles. At the gas station, you are pre-purchasing miles — this tank of gas will let me go 300 miles. With gas cars, the fuel is a separate purchase from the car. Better Place has done something no other electric car company has done — we've also separated the fuel from the car."

Mr. Granoff explains that, unlike other electric cars, which use built-in batteries that require the car to go idle while the battery is recharged, Better Place cars use a swappable battery.

Most electric cars juice up at electric "filling stations" when they're not in use. You have a charger at your home, a charger at your office, and you plug in whenever you're going to be parked for a while.

Better Place cars do that too.

But since most electric cars lose their charge after roughly 100 miles, that means you have to park for the hours it takes to fully recharge your battery — or you get stranded.

Better Place cars also offer that ballpark 100-mile range . . . but if you manage to discharge a Better Place car's battery, you just stop into a Better Place battery switch station, and in a matter of minutes, the spent battery is swapped out for one that's fully charged. It's the same process we've gotten used to for filling up our gas tanks, but one that is far greener.

This means a Better Place car offers nearly unlimited mobility, and that is a game changer in the move to an all-electric car infrastructure.

But let's back out a minute: why are electric cars an improvement over our current gas-powered automotive world?

As Mr. Granoff, whose Better Place office is based in New York, explains: "The reason that I and everyone else got into Better Place is because we want to end the monopoly of oil on transportation. The world consumes 85 million barrels of oil per day, and that's not going to go to zero overnight. But we can move our transportation off of oil, and we should do it for a lot of reasons."

Chief among them: the unfortunate truth is that, aside from the pollution generated by the millions of tailpipes out there, oil is owned and controlled by some really bad actors — nations whose interests run counter to ours, and, in a lot of cases, counter to their own people. Consider any of the wars the US has recently fought, and you don't have to look too far to find oil lurking there as a contributing factor.

Mr. Granoff takes another angle on the situation: "The US trade balance is a particular problem. Currently, our economy sends one billion dollars a day to oil producing countries. And it's growing. The rising price of oil means that every household in America is spending $4,000 more per year on oil than it did ten years ago. In an era when we need job growth and economic activity, that's $4,000 a year that each household isn't spending at the hardware store, at Starbucks, at the grocery store and so on . . ."

While the geopolitical and economic advantages of electric cars seem beyond reproach, some have criticized electric cars as a lateral move environmentally. After all, the electricity to run a car has to come from somewhere, and in the US, that largely means coal.

Mr. Granoff understands this point, but he says it's not a reason not to move to electrics. "Right now," says Mr. Granoff, "it is more efficient to generate electric power in one central location than it is to generate gas power in one million car engines. And by taking all those tailpipes off the road, you vastly improve air quality in cities. Eventually, as clean fuel alternatives like wind and solar proliferate, any environmental impact from electric car power will go away altogether."

In fact, says Mr. Granoff, the biggest obstacle to an electric car infrastructure has been two-fold: structural scale and political inertia. But Better Place has been able to overcome those factors by picking specific starting points and being donkey stubborn and putting their operations in place.

The result was Better Place's historic agreement with Israel to build the first nationwide electric car infrastructure.

The company began by installing 40 battery switch stations around the country, which will enable drivers to go, as Mr. Granoff says, "from Lebanon to Eilat and back again" without having to worry about running out of go-juice.

Additionally, the company is placing hundreds of charging stations at malls, hotels and office complexes around the country — wherever people park long enough to add a meaningful charge to their battery.

With the system now in place, the company is preparing to build out its car sales by the end of 2013. Better Place partners with Renault to provide the automobiles, which went on sale in 2011 in Israel. Mr. Granoff says sales of automobiles in Israel number about 200,000, and Better Place expects to capture 10% of that market by the end of 2013. That would make Better Place the biggest car seller in the country. And that's when one of the other benefits of electric cars really kicks in.

When electric cars charge up during peak times, they draw energy from a country's energy grid. But when an electric car is on a charger at off-peak times — say, when it's hooked up to a home charger at night — the car can actually sell electricity back to the grid. This creates a whole new market of electricity trading (as well as providing a security benefit; "You've got a big part of a nation's power supply distributed all over its area, stored in car batteries," Mr. Granoff says).

This becomes a big advantage to places like Denmark — Better Place's next target market, where it plans to put its cars on the road in the second quarter of 2012.

"In Denmark, the wind comes at night," says Mr. Granoff. "Formerly, this generated so much extra energy that Denmark was paying Germany to take it off their hands. Now, all that energy can go into electric cars. They keep their energy supply within their own borders."

After the Danes start firing up their Better Place cars, Australia will be right behind, followed by southern China. Mr. Granoff says Better Place hopes to own 10% of each market within two years of launch.

And as for the United States? Mr. Granoff becomes wistful. "The US is a different case. Because of subsidies that fossil fuel companies get and that alternatives don't, it costs about half as much to fill up a car here as it does in the rest of the world. We've come out with plans to put grids into California, along the whole Pacific coast, even a taxi fleet in San Francisco. They'd work. But without the kind of government subsidies we'd need to make ourselves cost-competitive with gas — subsidies the oil companies are getting — we can't feasibly enter the American market when there are so many other profitable places we can go."

True, electric cars like Chevy's Volt and Nissan's Leaf have received a fair amount of attention here. But Mr. Granoff says their impact is minimal. "Right now, hybrids and electrics are bought by ideological people, mostly on the coasts. They're not being bought as the result of an economic decision."

Put another way, electrics are not big sellers. According to the New York Times, G.M. has sold about 8,000 Volts since introducing it in late 2010. "This is in a market of 13 million cars," says Mr. Granoff.

America's reluctance to make a move to electric cars is deep-seated and complex. But one thing about it is clear: as other countries move to electric car systems, as they begin to reap the benefits of oil independence, cleaner air and a more self-contained transportation system, America will find itself at an increasing competitive disadvantage. So in my opinion, the longer we delay making the switch, the more we'll ultimately wish we had.

Because the good news is, the rest of the world is turning to electrics. And when a car becomes a best seller in one country, that tends to affect how car sales go in other countries, and though the oil companies might not want to hear it, it's clear what direction the world is heading. While embracing electric cars might require a bigger leap than going all-in on a new bulletin board, I'd wager it would change our lives even more.

UPDATE: *A Better Place goes bankrupt, but finds a bright new investor.* In May 2013 A Better Place filed for bankruptcy. Despite its innovative approach to ending oil's monopoly, the business model didn't work. In July, A Better Place was purchased by Sunrise, a new investment group headed by green-tech entrepreneur Yosef Abramowitz.

4. Desalting the Ocean

Yoram Oren's work is unnatural.

As a member of Ben-Gurion University's Department of Desalination & Water Treatment, Professor Oren's job is to make water turn against its very nature and go places it doesn't want to go.

"Nature seeks equilibrium," he says. "Desalination, separating the salt from sea water to make fresh water, is an act of overcoming what nature is seeking. It's not easy."

It's not easy, but as the Earth becomes more crowded there simply won't be enough naturally occurring fresh water to go around. We're going to have to get more of it, and the only feasible way to do this is to make it from sea water. Over the next 10, 20 or 50 years, desalination will become not only important but compulsory.

But while the rest of the world is just waking up to the idea that desalination will become a way of life in the years to come, desert countries have been relying on desalination for decades, because they've had to.

There are a number of ways to achieve desalination and they don't all require heavy equipment or tricked-out laboratories; you can do it crudely by boiling water on your stove top. But it wasn't until the 1960s that an Israeli scientist by the name of Sidney Loeb developed a new method of desalination that launched the process into the modern era.

Working at UCLA, Professor Loeb developed a semi-permeable membrane that made the desalination process known as reverse osmosis a practical and affordable way of making fresh water.

As described by Prof. Oren, reverse osmosis requires a change in the direction that salt water naturally seeks. If you'll allow an illustrative if inexact science lesson from a non-expert:

Imagine a bathtub. Now imagine stretching a piece of fabric across the middle of the tub. You fill one half of the tub with seawater, and one half with fresh water. The tendency will be for the fresh water to migrate across the fabric "filter" and mingle with the salt water, producing a solution (called brine) that's less salty than the seawater, but more salty than the fresh water. This is because nature likes to achieve equilibrium whenever it can — substances will move across areas of high and low concentration until uniformity has been achieved.

That process is called osmosis. Reverse osmosis does what the name implies — it turns osmosis on its head and creates conditions where two solutions will move away from equilibrium.

To achieve this, instead of using a "dumb" filter placed between two bodies of water, reverse osmosis uses a smarter semi-permeable membrane which allows only smaller water molecules to pass through. But a water molecule's instinct is to head into the salty area to help dilute it. To get it to go in the other direction, reverse osmosis applies energy to "push" salt water through the filter into the area of fresh water. This process extracts fresh water from the salt water, leaving the salt behind.

Scientists had been aware of reverse osmosis for centuries, but achieving it on a useful scale had escaped them, because forcing sea water to go "uphill" through existing membranes required a lot of energy — too much energy to be practical.

Prof. Loeb's breakthrough was the development of a new kind of semi-permeable membrane, and a new, energy-efficient way of pushing seawater through it.

The new methodology involves much in the way of advance

electro-chemical engineering, but suffice it to say that the method took off, and according to Prof. Oren, it's recognized around the world as the most advanced, efficient way of freshening seawater in use today. And we should hope that the trend toward Prof. Loeb's methodology continues.

"Around the world, water is still being desalinated by simple distillation, especially in oil-rich countries, which can afford the fuel to boil water," says Prof. Oren. "But they're changing, because water is going to be the big challenge of the next century."

Prof. Oren points out that in places like Saudi Arabia, and even in his own hometown south of Tel Aviv, desalinated water is the sole source of drinking water. Israel's Ashkelon plant is the world's largest reverse osmosis facility, producing 320,000 cubic meters of fresh water every day — meeting the needs of roughly 100,000 people.

That dependence is only going to grow, and even countries that are profligate with their resources are coming around to water treatment through reverse osmosis — especially since Prof. Oren and others in his field are starting to realize how much more reverse osmosis can achieve.

"The environment is telling us to find more efficient ways to treat not just saltwater, but all our wastewater, agricultural runoff, and municipal water," he says. "We're working on making membranes that are more and more sophisticated because now we have to protect them not only from salt, but from organic compounds, inorganic compounds, bacteria, viruses, proteins, sugars, all the stuff you find in the different kinds of waters we're treating."

Industrial-scale reverse osmosis is becoming the norm in places from Malaysia to Spain — which, according to Prof. Oren, seems to be in a friendly competition with Israel to build the world's largest desalination plant. Spain is also currently building 20 desalination plants, which will handle about one percent of the country's total water needs within just a few years.

Between the demands of population growth, increasing energy consumption and more people wanting a better way of life, conditions on Planet Earth are changing. Right now, the shape and tone of the lives we'll lead in the future are up for discussion. But some things are essential, and none of those is more vital than clean fresh water — and lots of it. While the rest of us come to terms with how we want to live life in the future, Prof. Oren and his colleagues are building on the work of Sidney Loeb to ensure that we'll have the water we need to live it.

5. Spinal Surgery Robots

If you were to get all your medical information from watching Mixed Martial Arts — where participants get pounded, kicked and punched for a living — you'd think that human beings are made of the toughest stuff in the universe.

And you wouldn't be far off. Pound for pound, human bone is stronger than steel.

But if you ask medical professionals who work with the human body when things go wrong, they'll tell you that along with being exceptionally resilient in the face of violent physical punishment, our bodies are also extremely vulnerable to the tiniest of threats.

And no one knows this better than spinal surgeons.

Patients undergo spinal surgery for a range of reasons — everything from the treatment of spinal cord tumors to the placement of screws in cracked or malformed vertebrae.

All forms of surgery entail risk. But spinal surgery leaves literally no margin for error. If incisions, instruments or implants miss their target by micrometers and damage the spinal cord, the result can be permanent nerve damage, paralysis or worse.

All from the tiniest little nick.

The trouble is that the human hand and the human eye have a limited capacity for precision — and the field of spinal surgery is where doctors bump up against that limit most calamitously.

Statistically, about 10% of spinal surgeries result in the need for revision — that is, they have to be done over. And between 3–5% of them result in permanent nerve damage.

According to Prof. Moshe Shoham, "that's remarkably good, but it's not good enough. We need to be accurate, period." Prof. Shoham has made it his business to increase the accuracy of spinal surgery procedures and bring about a significant improvement in their outcomes.

But if human surgeons have maxed out their accuracy, can those sobering statistics be improved upon — or even eliminated?

This is where the area of surgical robotics comes in.

Many analysts agree that surgical robotics — machines that enable surgeons to perform procedures more steadily and accurately than they could on their own — will be one of the hot growth areas for medicine in the decades to come. And nowhere will the reliability of robotic machines be more welcome than in the field of spinal surgery.

But whereas the medical establishment and those who observe it are just now catching on to the vast expansion of skills that robots can offer to spinal surgeons, Prof. Shoham has seen this coming for a while.

Specifically, since 1984.

That's when Prof. Shoham, a mechanical engineer who harbors a lifelong fascination with all things robotic, published a book about robotics and declared "the next revolution will be in personal robotics" — the area where robots have a direct effect on activities performed by people.

Prof. Shoham has applied his talents to aeronautics and studies of mechanical manipulation, but the medical field is where he's made his greatest contributions.

Prof. Shoham is the Founder and CTO of Caesarea-based Mazor Robotics, and is the technical and engineering mastermind behind one of its flagship products: the SpineAssist surgical robot.

If you're like me, when you think of a medical team turning to something called the SpineAssist robotic surgical assistant, you expect something like a sterilized Tin Man to come tottering out of an operating room closet with a tray of scalpels and a mop.

The reality is much less visually imposing but much more ingenious.

The SpineAssist is a cylindrical robot that looks a little like a soup can with a pea-shooter attached. But the technology it employs goes a bit further than that.

In the OR, the SpineAssist is mounted on a T-shaped frame that's "bolted" directly to a patient's back, by means of screws that are sunk into a patient's hip bones.

This means that the SpineAssist moves with the patient, as he breathes or makes minute muscle movements.

"If you can make the robot move with the patient, then the robot becomes part of the patient, and then you get the accuracy you need," says Prof. Shoham. "Though other companies make surgical robots, ours is the only one where the robot works as part of the patient. If you're not attached, you lose accuracy. When you're attached, you eliminate a big source of error."

The SpineAssist works in tandem with imaging and positioning software that allows for precise placement of a guidance tube — that pea shooter — above the patient's spine. The surgeon can then insert instruments through the tube and into the patient's back to achieve a level of accuracy that, until now, had simply not been possible.

How accurate is the SpineAssist?

"The robot has been used in over 2000 cases in the U.S., Germany, Russia and Israel, inserting more than 15,000 implants, such as orthopedic screws. So far there has been zero nerve damage, zero need for revision."

Zero.

Assume the low-end estimate: 3% of conventional spinal procedures result in some form of nerve damage. That means that, thanks to SpineAssist, there are 60 people out there without nerve damage who otherwise would have suffered some ill effect of their surgery.

When you consider that there are more than 250,000 back surgeries conducted in the US alone every year — and that this number will only rise as our population ages — you realize that the SpineAssist can protect thousands of people from the tragic consequences of spinal surgery that, statistically, they would have inevitably suffered.

Of course, as with any radical innovation, Prof. Shoham and his team at Mazor have run into some resistance as they attempt to deploy the SpineAssist in American ORs.

"Medical procedures have become routine, and for good reason — but that means there's resistance to change," Prof. Shoham explains. "I can understand why."

Still, Prof. Shoham is forthright in his conviction to put his robot to work for the benefit of spinal patients around the world. "I do think that, ultimately, if you're really good, you can convince people to make change, and I'm confident that we are going to do that, and that our robot will help to save and improve thousands of lives."

If focusing on the SpineAssist means the auto-mop robot will have to be put off for a few years, that's a trade-off I'm willing to make.

6. Pilotless Drones

In the movies, desperate citizens trapped in a burning high rise can count on Batman to swoop in and rescue them. But if Batman himself got trapped? He'd probably push a red button on his utility belt and summon The AirMULE — a pilotless, remote controlled "flying platform" that seems like it's right out of a comic book . . . except it's intended to serve in the most dire real-world circumstances.

Built by Yavne-based Urban Aeronautics (UA), the AirMULE is designed to enter environments that no other air vehicle can get to, from urban disasters and combat zones to treacherous mountaintop terrain. Its job is to ferry supplies to emergency workers, troops and civilians and extract casualties.

According to UA founder Dr. Rafi Yoeli, the AirMULE represents a leap forward in pilotless technology that will pay dividends in countless situations.

"The need became apparent when UA was working on a larger twin-engine manned air vehicle," Dr. Yoeli says. "First we heard from the US Marines that there was a pressing need for unmanned logistic supply and casualty evacuation. Then we heard it from the Israeli forces. Then we heard it from NATO. They were all saying the same thing: obstructed terrain, hostile environments, and a pressing need to get materials in and people out. It was obvious to us that our technology had a huge advantage over anything else out there, and that people would really benefit from it."

Listening to Dr. Yoeli describe what makes the AirMULE tick — and what technological hurdles his company had to clear to make it operational — it quickly becomes clear that you're in the company of someone who's passionate about the field of aeronautics. Which makes sense, as Dr. Yoeli is a Boeing alum who also serves as a reserve officer in the Israeli Air Force.

When Dr. Yoeli gets warmed up about the AirMULE, the talk quickly turns to matters of "vane control" and "adaptable duct airflow" and "six degrees of freedom." The tech speak is impressive and indicative of deep expertise — but for the average person on the street, what it all adds up to is a bit of an engineering miracle.

The AirMULE is driven by "internal rotors" — essentially large, enclosed fans that allow this flying rescue-mobile to go where helicopters, with their large, exposed blades, simply can't.

Though the project is rooted in military applications, its relevance to everyday situations was immediately clear to Dr. Yoeli and his team. That's why they're eager to move the project past the prototype stage and into full mission demonstrations.

That's something that communities around the world should hope for, because when you get a full grasp of the AirMULE's capabilities, the ways it can help become readily apparent:

- A wicked rush hour car wreck in a busy intersection causes gridlock in all directions. Ambulances can't get through. Fire engines, ditto. Close quarters and limited landing space mean helicopters are not an option. But AirMULE can slip into the tightest urban spaces, deploy a Medivac team and get the victims out while traffic is still frozen in place . . .

- A tornado levels wide swath across multiple states. Personnel are stretched thin. AirMULEs can be sent into remote areas with supplies . . . and because they're pilotless, an operator can then leave one AirMULE on the ground and deploy another one to a different area, saving time that would be spent on return flights for piloted aircraft . . .

The scenarios where AirMULE can help out are numerous — and

likely to multiply as budgets for civic services shrink in a world of proliferating natural and geo-political threats.

It might not stop trouble in the first place, but AirMULE represents a dramatic advance for the ability of people to help one another out when the worst happens. It's smart, it's elegant, it makes sense and anyone who sees it — and that includes Batman — is going to say: I want one in my town.

Though Batman might also add: does it come in black?

7. Diagnosing Sleep Apnea

Dov Rubin is not your mother.

But he does want to make sure you're getting enough sleep.

He has good reason for concern.

According to the National Sleep Foundation, about one in five Americans report that they average fewer than six hours of sleep per night. Everything from stress and diet to bad shades and late night TV-watching can contribute to insufficient sleep . . . but one of the most notorious and elusive culprits is a medical condition: sleep apnea.

Sleep apnea is a disorder in which a person in deep sleep actually stops breathing. This happens because the sleeper is so relaxed that the muscles around her airway go limp and "collapse," causing the airway to be reduced or entirely cut off.

The sleeper then partially wakes up — coming up from restorative deep sleep into the lighter REM sleep — until breathing is restored. Then she falls back into deeper sleep and the whole process starts all over again.

These breathing gaps can last for up to several minutes at a time, which sounds brutal. Even worse, the gaps can be really short — a matter of seconds — which means that the waking can take place several dozen times per minute.

Most unnerving to my mind: typically, people with this condition aren't even aware it's happening. All they know is they wake up after eight hours feeling like they've slept for something far less than that. In fact, it often takes a very alert spouse or partner to notice how poorly his bedmate is doing in the sleep department; it's not always obvious.

The truly unfortunate part is that, once diagnosed, sleep apnea is fairly easy to treat. Simple fixes from losing weight and keeping a regular sleep schedule to using a second pillow can help most sufferers. More severe cases can be treated with sleep masks or, in extreme cases, simple surgical procedures.

Medical science has been sensitive to the problem of sleep apnea for years and has developed a protocol for identifying it. That approach can be described as clunkily effective . . . but the problem is that it's also cumbersome, with significant room for improvement.

Dr. Dov Rubin wants his company to move into that room.

Dr. Rubin is President and CEO of Itamar Medical, a Caesarea-based medical device company that has devised and deployed the WatchPAT, an extremely nifty little device that represents a 180-degree pivot — and dramatic upgrade — from the way sleep apnea has traditionally been diagnosed.

"Ever since sleep apnea was identified as a problem, it's been diagnosed in a laboratory setting," Dr. Rubin explains.

This makes sense because analyzing sleep patterns to determine how deep you're sleeping and when you're waking up has always meant collecting information about your heart rate, your breathing rate, the amount of air you're taking in, your sleep state, your brain waves — in other words, lots of data.

The process of harvesting this information usually involves having patients spend the night sleeping in a lab bed, hooked up to a noodle salad of sensors, all under the watchful eye of lab technicians.

"It's definitely possible to gather useful information that way," says Dr. Rubin. "But there are also a number of problems that go with the laboratory approach that simply can't be avoided . . . and that are actually self-defeating."

Dr. Rubin cites two major shortcomings of the laboratory approach.

"First, half the population is never going to set foot in a sleep lab. Whether it's right or wrong, there has historically been major resistance on the part of women to putting themselves in a position where strangers with cameras are going to watch them sleep, and frankly I don't blame them."

Second, there's what Dr. Rubin describes as the First Night Effect.

"Imagine how comfortable you'd be if you stepped into a sleep lab, got all wired up by a bunch of folks in lab coats, they tuck you into an unfamiliar bed, and then they say 'now have a comfortable and representative night's sleep.' You know that's not going to happen. So frequently they have to keep you over for a second night to get a real picture of what you're doing when you sleep."

Sleep testers have attempted to work around the alienating sleep lab environment — but those fixes have usually involved replicating the lab experience at home. So you might feel a little more relaxed in your own bed — but no matter where you are, it's hard to be fully at ease with sensors stuck to your head, a belt around your chest and an air tube in your nose.

Hence WatchPAT.

The WatchPAT is basically a sleep lab that you wear on your wrist (that's the "watch" part) while you sleep; the device attaches to your fingertips and measures signals your circulatory and respiratory system generates during sleep (PAT = peripheral arterial tone), and translates them into a wealth of information about how well you're sleeping and what you're up to while you're doing it.

The WatchPAT traces its origins to the operating room. According to Dr. Rubin: "The story goes that sometime in the mid 1980s, a surgeon noticed that as he was squeezing one of the arterial nerves, he was getting a reaction from the fingertips. This meant there was a correlation between the fingertips and the arteries."

This was the flash of insight that eventually led to the development of the WatchPAT. (By the way, investigations in this area also led to a Nobel Prize in Medicine).

One reason the WatchPAT is noteworthy is that it gets at a whole new range of information about sleep.

"The device provides a window of information never before available in such a convenient and non-invasive package," says Dr. Rubin. "We are monitoring not just sleep but the entire autonomic nervous system" — the part that works without you even being aware of it.

At the end of a night's sleep, the WatchPAT generates a report detailing how long you slept, how long you spent in each of the four stages of sleep, whether you snored, when you snored, how snoring correlated with your sleep stage, your level of blood oxygen during each stage . . . the list goes on. All of this can tell you whether your sleep is actually delivering the rest you need — or whether it's making you more tired than you ought to be.

But when you start thinking of how this information can be applied to the general population, you start to see how wide an effect the WatchPAT can have on the way we all live our everyday lives.

First, there's the extent of the problem. Dr. Rubin estimates that somewhere on the order of 30 percent of the population is suffering from some degree of undiagnosed sleep apnea.

That sounds like a lot of zombified people walking around out there. But that's just the point: a lot of them aren't simply walking around.

Bus drivers, airplane pilots, cops, sleep lab technicians . . . lots of people with our lives and welfare in their hands are going about their business in a state of profound exhaustion.

WatchPAT can help. Dr. Rubin says that as Itamar gets further into its work with the WatchPAT, it is discovering more and more applications that provide ground-level benefits to the population as a whole.

Take the case of long-haul truck driver exhaustion — something that has been on the rise in recent years. Many states are now adding laws requiring that truck drivers get an annual test for sleep apnea.

But with 3–4 million truckers in America, there simply aren't enough sleep labs to handle the load — even if the labs delivered perfectly efficient results every time — not to mention the difficulty in asking these folks to take four days out of their schedule to travel to and get tested at a lab.

But a driver can simply take a WatchPAT home — or even into the back of the truck cab (which is often tricked out like a Holiday Inn). By using a bracelet with an embedded chip that ensures the WatchPAT will be used only by the intended patient, the driver can take the test at her convenience without losing precious driving time.

The immediate result: accurate and timely information about the driver's condition. The longer range result: fewer exhausted drivers getting involved in fewer accidents and a potentially drastic reduction in road fatalities.

Once you grasp how widespread the problem of sleep apnea is, you can see the ripple effects of the WatchPAT. Along with American medical practices, clinics in Japan, South Korea, India, China and Scandinavia are using WatchPAT.

And along with these individual practices, lot of organizations are clueing in. So while medical organizations like Kaiser of Northern California and the Veterans Administration have been major participants in the 500,000-plus sleep tests the WatchPAT has conducted around the world, other outfits like Union Pacific railways are starting to come on board as users of the device. As I'm typing this, worldwide sales of the WatchPAT have passed the 3,500 mark. (To put that number in perspective, consider that the US is home to 1,400 accredited sleep labs; that's 2.5 WatchPATs for every lab in America).

As exciting as this is, Dr. Rubin gets even more exercised when discussing the future of the device.

"We are just starting to realize exactly how much information we are getting from this little fingertip signal," he enthuses. "Yes, sleep apnea is a cause of conditions like obesity and cardiovascular disease . . . so if you bring sleep apnea down you can really curtail those conditions.

"But we're also learning how to apply this signal to the diagnosis of depression and atherosclerosis, erectile dysfunction and preeclampsia and Alzheimer's . . . and even to the development of a foolproof lie detector . . . we are just starting to get a handle on the richness of this information."

Managing its success is a big challenge for the company, especially as it attempts to mine its "non-invasive window of information" for every iteration it can think of.

But the effort is well worth it for Dr. Rubin, who says he thinks every day of a dentist he knew who was complaining of chronic exhaustion. Dr. Rubin gave him a WatchPAT test, and discovered that while he was asleep, the dentist's breathing was stopping an astonishing 78 times per hour — about once every 45 seconds — for more than 20 seconds at a time. The man was literally choking more than he was breathing.

Once diagnosed, the dentist took some simple steps and basically became a new man.

"He was very grateful, but the real reward comes from his wife," Dr. Rubin says. "Every time I see her, she thanks me for saving her husband's life."

8. Safer Driving with Mobileye

There are those who argue that we live in a world of relentless fear, strife, conflict and despair, and that we are a species whose members are internally wired not to get along with one another.

Whenever I'm confronted by a proponent of this grim outlook, I counter with two pieces of evidence that prove that humanity is not too far gone, that there is within the human animal a wellspring of goodness and the ability to co-exist.

Exhibit A: bakeries. In the face of the daily micro and macro discord that obtains across Planet Earth, certain people — from all over the world — still get up in the morning and dedicate themselves to the creation of dainty cakes and crumbly cookies and assorted treats, going so far as to decorate them with frosting flowers and rainbow sprinkles. This fills me with hope.

Exhibit B: driving. In the abstract, if you were to describe to me the act of driving as it's commonly practiced — millions of cars hurtling past each other on highways and weaving amongst one another in jam-packed cities — it would come in very low on my list of things I'd expect humans to do well. Yet most people can actually stay on their side of the line most of the time. I find this amazing, and cause for pride in people.

And yet . . . according to the National Highway Traffic Safety Administration, about 43,000 people are killed in auto accidents every year in the US alone. There's no doubt — we have room for improvement.

Amnon Shashua has moved into that room.

Prof. Shashua is Chairman and CTO of Mobileye, a company based in Jerusalem that develops systems that cars use to keep drivers safe and reduce traffic accidents — all through information gathered by a single camera mounted on the windshield.

"We use a forward-facing camera mounted on the windshield, and the video from the camera is processed by our technology to produce driving systems that directly aid a driver," says Prof. Shashua. "In short, we make systems that let cars see."

The result goes far beyond the simple backup camera that helps you parallel park.

Mobileye's Advanced Driver Assistance Systems (ADAS) is an outgrowth of Prof. Shashua's work in the field of Computer Vision and Machine Learning — and its capabilities are astonishing.

- Mobileye can warn you if you're drifting out of your lane.
- Mobileye can track the vehicle in front of you and calculate the time to a collision.
- Mobileye can monitor the sidewalk and distinguish between non-threatening objects (a mailbox) and something that might pose a risk (a pedestrian).
- Mobileye can interpret the body language of a pedestrian — a turn of the torso, a dip of the shoulder — to determine if the pedestrian is about to cross the street.
- Mobileye can read street signs, detect bicycles, and decide whether you should be using your low beams or your high beams.

As Mobileye sees and senses and interprets all this information, it uses a range of signals to pass all the information on to the driver.

"Mobileye is customizable," says Prof. Shashua, "so it can let you know you're leaving your lane by vibrating your steering wheel, or sounding an alarm. It can respond to an oncoming accident by tightening your seat belt and priming your brakes, or it can go so far as to actually brake your vehicle. In essence, the system's job is to be paranoid so you don't have to be."

When you wrap your head around all of Mobileye's capabilities, you understand that it represents the first step toward the future of driving. And you also realize the amount of brainwork that went into it.

"The first challenge is a common challenge: cost," says Prof. Shashua. "Humans see with two eyes — bicameral vision, which provides depth perception. But achieving bicameral vision on a car is very different than achieving it on a human."

Prof. Shashua explains that using two cameras on a car would add significantly to the cost of the system, especially once you factor in not only the cost of an additional camera, but also the additional processing power and "eyeprint" the second camera would have on the car's available real estate.

"You need to have two cameras a certain distance apart to achieve depth perception at a useful distance down the road," says Prof. Shashua, "and the fact is that there aren't two points on a car that are far enough apart."

The solution was to use one camera to interpret all the information that's out there on the road. But this raised its own set of problems — primarily, the fact that a single camera capable of gathering visual information and processing it in an appropriate way didn't exist.

In fact, says Prof. Shashua, "the idea was laughable."

Prof. Shashua and his team went ahead anyway. From somewhere near scratch, they developed the chip and the software that make the Mobileye work. Their work represented a breakthrough in the world of computer vision. And automakers responded — albeit with skepticism at first.

"Car makers treated our proposal with disbelief," Prof. Shashua says. "How can you have a system that can interpret information with a single camera, and make it affordable? But we convinced them to let us test it." And based on the test results, the automakers became believers.

Mobileye first struck deals with BMW, Volvo and GM. By 2007, they had systems in place with the majority of carmakers. In more recent years, they've gone global, closing deals with Hyundai, Ford and Opel.

Carmakers aren't the only ones filling up on Mobileye's technology; once they heard the pitch, Goldman Sachs invested $100 million to help get the company on the road.

To date, Mobileye has sold 500,000 factory-installed units and another 60,000 after market systems. The company owns about 80% of the market for automotive smart cameras.

But according to Prof. Shashua, that's changing. "In 2009, our system would cost $3000, and so they were available only to those who could pay that kind of money. Two years later, it cost $250. Before very long, these systems are going to be standard equipment on all models of all vehicles. They'll simply be built in."

And that would unquestionably be a good thing. In a 2009 study, the Federal Motor Carrier Safety Administration determined that forward collision warning systems and lane departure systems could reduce accidents involving trucks by anywhere from 13,000 to 25,000 per year.

Buried in that statistic is an interesting nugget: Mobileye not only makes driving safer for those who use it; by reducing accidents, it also makes driving safer for everyone else on the road.

So as Mobileye continues to develop its driver assistance products, and as it helps to spawn the next generation of smarter, nimbler vehicles, we all have a stake in its success.

Vehicle makers can offer improved safety as a selling point.

Insurers like the reduction in claims that goes with the reduction in accidents.

And motorists like me really appreciate the idea of getting where we're going in one piece — whether it's the bakery, or baking class, or anywhere in between.

UPDATE: *Mobileye gets a major cash investment.*
In August 2013, after a $400 million cash injection from Goldman Sachs and others, Mobileye was valued at just over $1 billion.

9. Clean Fish Farming

There are a lot of fish in the sea . . .

This little axiom has given comfort to lovelorn adolescents for generations. It's also been the long-term mantra of the world's industrial fishing community.

But if mankind continues to pull fish from the world's oceans at our present pace, the lonely-hearts crowd is going to need to look for a new morale-boosting analogy. And the world's fishermen are going to have to look for a new line of work.

Estimates vary, but the consensus seems to be that unless drastic measures are taken to reduce the catch of the world's fishing fleets, our oceans will be effectively empty of most fish stocks by 2050.

Wow. Empty.

This would be a global disaster on many levels. Environmental: people are land creatures, but this is a water planet, and if you take fish out of the equation, the ripple effect would fundamentally alter everything about the ecology of this place.

Cultural: Most of the world's coastal communities are linked either directly or indirectly to the fishing industry. If fishing goes, so does the identity of these places — and the billions who live there.

Humanitarian: Fish are food. According to a 2011 report by the Food and Agriculture Organization, people worldwide get about 16% of their protein from fish. That's an average, so obviously a lot of people rely very heavily on fish for sustenance. Take away the fish, and you're going to put those people in an extremely desperate situation. And world population is increasing, so we're going to be putting more and more pressure on the world's food supplies in the years to come . . . and it's not just "emerging" countries that are straining the system.

The USA is the world's third largest seafood market after China and Japan; the National Fisheries Marine Service reports that, as of 2008, US citizens were consuming about 16 pounds of seafood per person per year.

And along with everything else, the disappearance of the world's fish would be a moral catastrophe. We humans are aware of what we're doing to the world's fish. If we continue to do it without taking any counter-measures — condemning millions to starvation — we'll have only ourselves to blame as we live with consequences we saw coming a long way off.

But Dotan Bar-Noy and the company he heads are not about to let that happen.

Mr. Bar-Noy is into technology . . . but not of the X-box variety. He works with technology that benefits people at the most fundamental human level.

"An iPhone is nice to have," he says. "But you need to eat. And you can't eat an iPhone."

Mr. Bar-Noy is CEO of Kadima-based GROW FISH ANYWHERE (GFA), a company that enables people to raise pretty much any kind of fish anywhere on the planet with absolutely no environmental impact.

"As the world's population increases, we're not going to be able to satisfy everyone's protein needs with meat," Mr. Bar-Noy says. "Not because there's not enough meat, but because there's not enough water."

It turns out that meat is very water intensive. It typically requires about 500 gallons of water to produce a pound of beef protein. Other common sources aren't much better: a pound of pork protein requires 100 gallons of water, and a pound of chicken protein requires 80 gallons of water.

"Ironically," says Mr. Bar-Noy, "given the shortages of water that the world will be facing, it is fish that has the greatest potential as a food crop."

Fish farming is not new — the practice of "aquaculture" has been around as long as people have been looking to the sea for sustenance. But in recent years, aquaculture has been conducted at an industrial level that, according to Mr. Bar-Noy, raises as many problems as it solves.

"Fish farming traditionally has happened in one of two ways," says Mr. Bar-Noy. "The first way is to raise fish in the ocean, right by the shore, in large nets. You're relying on the ocean to circulate water and provide a clean environment for the fish. But with the amount of fish you need to make a difference, you don't get enough ocean movement to keep the water clean. The result is very dirty water."

According to Mr. Bar-Noy, industrial fish farmers have tried to solve the dirty water problem by moving their farming operations to land-based tank systems near the shore.

Raising fish in tanks has traditionally reduced the problems of fish swimming in polluted water — ammonia and nitrogen are the major culprits — but Mr. Bar-Noy says that you still produce a major environmental problem for the sea.

"If you have one fish in an aquarium, you are okay," he explains. "If you add more and more fish, you need to replace some of the water every one or two weeks, say. With land-based commercial fish operations, when you pack two to five kilos of fish per liter of water, you need to change out a lot of water to keep the system clean."

Land-based operations solve this by pumping seawater through the tanks. This keeps the tanks clean — but it ends up pumping a lot of dirty water just offshore.

So there's the pollution problem all over again.

GFA was founded with the goal of doing things differently.

As Mr. Bar-Noy puts it: "We took the approach of looking at the problems aquaculture raises, and asking: What is the real goal here. It should be: how do you make a land-based system that is closed, that doesn't need to change water, and that can be operated wherever the need for food exists."

The result was GFA, an aquaculture system that produces zero discharge of water and zero environmental impact, that can grow just about any type of fish, and that can work literally anywhere on the planet.

The system is clean, green, sustainable and provides an abundant food source for a planet that is getting smaller and more crowded.

As with many game-changing technologies, the result is very simple — fish swimming around in tubs of clean water — but getting there required some highly singular thinking.

The main problem with closed systems is that nitrogen buildup in the water quickly makes the environment toxic for the fish.

The breakthrough idea came with the realization that using bacterial filtration — essentially feeding the wastes generated by the fish to a community of anaerobic bacteria, which consume the noxious compounds in the water — would enable the system to eliminate that excess nitrogen. With that filter in place, you can circulate the same water over and over and keep it clean indefinitely.

Voila! A totally closed system that produces one pound of fish protein using only about two gallons of water. And because you don't need to pump large quantities of water through it, you can put this system literally anywhere. GFA systems are designed to be modular — each one can produce between 30 to 70 metric tons of fish annually, and there's no theoretical limit to the number of units that can be operated together. To put this capacity in perspective, consider that an annual production of 1000 metric tons of fish can feed about 100,000 people per year.

GFA has systems operating in Israel and in New York State with more to come.

And Mr. Bar-Noy's vision for the future of the company is literally that: a visual image that lays out his intentions for GFA. "When you travel by train or car in the US, you see farms. In the future, where you see those farms, you will be seeing warehouses growing clean fresh fish."

And here's what I particularly like about GFA's goal. In addition to supplying a vital source of protein and feeding a hungry planet, the system also provides a way of life.

"Operating a GFA system is not entry level," Says Mr. Bar-Noy. "It's easy if you know what you're doing . . . but you do have to know what you're doing. It's all based on science, a lot of data, and you have to have a certain fund of knowledge to do it right."

In the same way that you can't be a successful farmer without knowing a lot about crops and the weather and soil chemistry and the way commodities markets work, growing fish successfully requires knowing fish biology and behavior as well as economics and marketing . . . in other words, fish farmers have to treat fish farming like a real business. These operations will call for smart, driven people to manage them. And they'll produce long-term, sustainable, challenging jobs in an area with the potential for huge, clean growth.

So: food for a hungry planet, jobs for eager workers, careers for ambitious business owners, and no ecological impact.

All in all, a good day's fishing.

10. The PillCam

Certain subjects have long proved problematic for photographers. Bigfoot. A baby sitting still. An internal view of the entire length of the small intestine.

Granted, it's possible that this last example isn't one that the pros have traditionally griped about while sitting around at the photographers' bar. But according to Homi Shamir, CEO of Given Imaging, since the advent of medical photography, the human small intestine was a really elusive fish — notoriously difficult to visualize without causing subjects an inordinate amount of discomfort and inconvenience.

"The small intestine is 21, 22, 23 feet long," says Mr. Shamir. "It's curvy and winding and folds back on itself over and over again. It's really difficult terrain."

Endoscopy — the practice of using tools to see inside the body — has been around since the early 1800s. Technological advances in the 1950s enabled doctors to insert cameras and lights inside of a patient's gastro-intestinal tract to screen for ulcers, polyps, lacerations and the like.

But there was absolutely nothing pleasant about the procedure — especially in the case of the small intestine.

"The length and architecture of the small intestine mean that it's almost impossible to see the whole length in one procedure with a typical endoscope," says Mr. Shamir. "You could put a scope in the top and see that half, and then in the bottom and see that half. But getting the whole thing at once — forget it."

The problems with this multi-step approach are just what you'd think: discomfort and inconvenience, giving traditional endoscopy an unpleasantness factor that is off the charts.

But things began to change in the mid-1990s.

It was then that Given Imaging's founder, Gaby Idan, was approached by a surgeon friend who lamented his inability to get a good, uninterrupted look inside his patients' small intestines. Not only were the pictures only so-so, but the nastiness of the procedure kept people from signing up for it in the first place.

At the same time, great advances were being made in the fields of miniature cameras, wireless communication and battery power.

The need and the technology came together in what was to become the breakthrough invention for Yoqneam-based Given. In 1998, the company introduced its signature device, the PillCam.

The PillCam is like an edible TV studio you can hold between your fingers. Elegantly simple on the outside, the PillCam is a feat of mini-engineering on the inside, combining camera, lights and wireless communication in a container no bigger than a fish oil capsule.

The result is an entirely different approach to medical imaging of the GI tract. And not just different.

"Compared to the old way of doing things, the experience is amazing," says Mr. Shamir.

Step One: The patient begins by strapping a fanny pack-size wireless data recorder around her waist. Then, Step Two, she swallows the PillCam — and as far as she knows, that's it.

Meanwhile, inside the patient's body, the PillCam makes an 8-hour journey through her digestive tract, snapping on the order of 60,000 pictures along the way.

The patient can carry on close to normal activities while all this is happening, and after the patient feels the capsule pass out of her body, she then returns the data recorder to the doctor.

The doctor downloads the images into a proprietary software program and examines them for indications of abnormalities.

With the PillCam, doctors have something they've never had before: a clear, continuous view of the small bowel, delivered with nearly non-existent inconvenience to the patient.

Itamar states that more than one million patients have been treated with the PillCam; it has been so successful — both in the USA and around the world — that the PillCam is now considered the gold standard for imaging of the small bowel. When a patient in pretty much any developed nation needs to undergo imaging of the small bowel, he's given a PillCam.

The company's future looks bright: Mr. Shamir says that it took Given Imaging nine years to sell its first million PillCams. It will take four years to reach the next million, and two years for the million after that.

The success of the PillCam has led Given to set its sights on several targets for the near future. Mr. Shamir says the company is working on devices for imaging the colon and the stomach. And the next generation of small bowel PillCams is not far behind.

Meanwhile, the company is working to develop capsules that can be steered by an operator. Right now, the PillCam is propelled through the body by the muscles of the digestive system. But Given is designing a capsule that can be "driven" by means of a magnet outside the body.

"Once we have independent movement," says Mr. Shamir, "we can take the camera from top to bottom, hover at certain spots, go back, look at things with a level of scrutiny that was never possible before."

After that, we get into truly futuristic territory. Given is working on tiny devices that can operate therapeutically in the body — conducting tests, taking tissue samples, delivering medicine.

As with any leading-edge technology, Mr. Shamir says that the path to the next generation of therapeutic robots won't be completely free of twists and bumps. But that's the nature of the field.

"A lot of people see a river and immediately want to build a huge bridge, with lots of infrastructure and expense," he says. "Our approach is, let's first see if there's something worthwhile on the other side. So we jump in the river and try to get across any way we can. If we like what we see over there, then we build the bridge. We might get a little wet, but at least we know where we're going."

It might not be the most common method — but this exploratory tack is the approach of a company that is determined to do a lot of hard work to make things a little bit easier for the rest of us.

UPDATE. *Success! The FDA approves the next generation PillCam.*
In August 2013 the US FDA said yes to the new-improved Pill-Cam SB 3, with superior video and image resolution for even better patient results. The PillCam now has over 2 million uses worldwide.

11. Floating Solar Panels

Yossi Fisher is looking to avoid a fight.

Mr. Fisher is co-founder and CEO of Solaris Synergy. Solaris is a solar energy company that did a very smart thing. It looked at a problem that has been dogging solar energy since its earliest days, and decided it didn't want any part of it. Instead, Solaris has taken a different route that has enabled the company to provide what might be the nimblest and most efficient solar energy systems available today.

Let's step out a moment and review a challenge that solar companies have historically struggled to resolve.

It seems that, ever since solar energy became "scalable" — that is, something that could work for a lot of people — there's been a pitched battle between solar energy companies, land owners, communities, and environmentalists.

The problem is that solar energy is generated with solar panels. And solar panels take up land. In fact, it turns out that producing traditional solar energy with enough wattage to make a dent in the power usage of real life communities means using a lot of land.

There are a lot of people who think covering the land with vast solar arrays isn't such a good idea. These include tourist agencies, builders, farmers, and even environmentalists, who aren't so enthralled with the idea of installing acres of solar panels and casting vast swaths of land into shadow.

"Solar panels have the potential to destroy nature," says Mr. Fisher, pointing out the irony of doing environmental harm with a technology that is meant to help the environment.

Mr. Fisher says a big part of this conflict lies in the fact that the PhotoVoltaic (PV) cells found in standard solar panels are only so-so in the efficiency department. To generate electricity at an industrial level, you need to deal out a lot of panels.

Mr. Fisher says that the main response to the efficiency question has been CPV — or Concentrated PhotoVoltaic — solar panels. This type of panel uses mirrors to concentrate sunlight onto a smaller array of solar cells. "It's a good idea," says Mr. Fisher, "but it has drawbacks. It throws a lot of heat onto the PV cells, which rapidly lose efficiency as they heat up. Sunny places are hot. And if you heat up solar cells, they lose efficiency. It's a paradox. When the sun is out, efficiency goes down."

Developers have tried to use different materials in CPV — the heat resistant substances used on spacecraft, for example. They work — but they are prohibitively expensive.

Another problem with CPV is that you need the mirrors that focus the sunlight onto the solar cells to track the sun precisely. If you lose focus, you lose the efficiency of the system — and any gust of wind that comes along can cause your system to lose focus. The solution to this problem has been to mount solar arrays on mechanical tracking systems. They're accurate — but, again, they're costly.

Mr. Fisher and his Solaris partners, a trio of physicists and engineers, looked at these problems and the solutions that hadn't proved up to the task — everything from using heat-resistant materials to clunky mechanical sun-tracking systems — and they knew they wanted no part of the land use argle bargle.

Yet they also knew that solar energy is a field with nearly limitless potential. Their eureka moment came when they realized that an answer was out there — and it wasn't on land at all.

Mr. Fisher and his partners realized that there are plenty of unused acres around the world where solar systems can be deployed: the planet is home to millions of small, medium and large bodies of water that are perfectly suited to host solar arrays. Installations from water treatment facilities to utilities, fish farms to reservoirs maintain bodies of water that can do double duty as solar sites.

But Solaris realized that while simply moving existing solar collection systems onto a body of water might solve the land-use problem, it did nothing to resolve the other problems associated with CPV technology.

"Cost is everything," says Mr. Fisher. "We wanted to be able to work in small areas, so we knew that CPV was the way we wanted to go. But you still have the problems of cooling the system, the loss of efficiency, the need to track the sun . . . But water provided the answer there as well."

Solaris completed their breakthrough by developing the world's first water-borne CPV solar array — a scalable system of panels that actually floats on the surface of a body of water.

While others have attempted to mount traditional panels over bodies of water — mostly as part of existing solar arrays that encroached onto wet ground — that still meant driving stanchions and pilings into the earth, just as you would on dry land.

Solaris panels are effectively untethered. They just float right on the water, as calmly as an inflatable raft in a back yard pool, and do their thing.

Aside from entirely mooting out the land use argument, floating a solar array on water presents a bunch of elegant advantages to solar users.

Not only does a floating system mean that you can track the sun frictionlessly, it eliminates the need for a heavy and expensive mechanical tracking system. Solaris systems also solve the problem of maintaining the temperature of its CPV cells by using the water they're floating on as a coolant.

Solaris systems can take their own temperature, and compare it to the temperature of the water, and regulate themselves so that they're always operating at peak efficiency.

Additionally, to make them as light as possible, Solaris systems are built with PVC tubing — they'll never corrode, and they're much cheaper than land-based systems, which by necessity are material intensive and subject to the ravages of the elements and time.

The question, of course, is: are there enough small bodies of water to make a dent in the planet's emerging need for solar energy?

According to Solaris, in Israel alone, there are over 400 relevant reservoirs spread throughout the country, with an average peak power generation capacity of 2 megawatts each. Counting on my fingers, I compute that out as 800 megawatts of peak power — enough to power about 800,000 homes. Mr. Fisher says that, give or take a megawatt or two, 800 megawatts is about what Israel has set as its solar energy target.

Solaris is saying that not only is Israel's goal for solar energy achievable, it's achievable without putting a single solar panel on land.

And that's just Israel. In the wake of successful pilot programs in the Negev desert and in France, Solaris is poised to enter markets around the world, with systems ranging from a small 50 kilowatt rig that could generate enough juice for, say, a farmer to run his operation, to entire power stations that can provide energy to villages and small cities.

Because while Solaris will allow any number of entities to harness the sun, another goal of Mr. Fisher and his colleagues is to provide power to individuals and municipalities who had previously gone without.

"We are working to bring a number of systems online in the USA," says Mr. Fisher, "but we're also looking to expand into Africa, India, and China. The hope is not to sell to a grid, but to sell to an actual village, at prices they can afford, and bring them the benefits of clean, reliable energy."

A clean system; a low-impact system; a system that resolves a long-standing, seemingly intractable conflict within the alternative energy community; a system that can mean energy independence to millions who have never known it. All of it coming in the wake of a single, simple, yet revolutionary idea.

It took people from a desert country to think of turning to the water to collect the energy of the sun. But when they did, they took a big step toward empowering us all.

12. Preventing Sudden Infant Death

Every year in the United States, 3,500 infants die of the utterly mysterious condition known as Sudden Infant Death Syndrome (SIDS). In the category of grim statistics, I find that one untoppable.

SIDS is the sudden and unexplained death of an infant who is less than one year old. It generally occurs when the baby is asleep. It's believed to involve a buildup of carbon dioxide in a baby's system and a baby's inability to wake herself up, but the medical profession is at a loss to state definitively what causes SIDS.

So far, the best that doctors have come up with is a useful but incomplete set of Dos and Donts (e.g., always put a baby to sleep on her back; do not put heavy bedding in a crib with a sleeping baby.) But while medical science continues to investigate the causes of SIDS, an Israeli company — Rishon LeZion-based Hisense, LTD — has deployed a device designed to sound an alarm when it might be happening.

And in one of those ironies of fate that turn up regularly in the world of inventions, the development of the BabySense Respiratory Movement Monitor started with an attempt to develop a monitor for the elderly.

"My father Haim invented the monitor, originally, you could say, for the elderly," says Hisense VP of Marketing & Sales Yaniv Shtalryd. "He had a doctor friend who approached my father and described to him a situation with the elderly that he said needed fixing."

Mr. Shtalryd says that the friend of his father's complained that elderly people are monitored by all kinds of high-tech, sophisticated equipment while they're in the hospital, but when they're discharged, most of them have to make do with rudimentary family care. The suspicion was that this situation plays a part in the high rate at which discharged patients lose their lives in their first night or two after a hospital stay.

Haim Shtalryd — an engineer by trade — set out to create a simple monitor for home care of the elderly, but he changed gears when he learned two things: one, there wasn't a lot of demand for a monitor for the elderly, and two, SIDS is the number one cause of death in infants under one year of age.

According to Yaniv, Haim looked at the existing baby monitors on the market and knew he could do better. "Other monitors at the time required the placement of sensors on the baby, which not only disturbed the baby, but posed problems of their own, like putting the baby in the crib with tangled wires and electrical current and so on."

In 1991, Haim Shtalryd developed a monitoring system that is completely hands-off. The BabySense monitor uses two sensor pads that go under a crib mattress. They communicate wirelessly with a monitor that mounts on the crib rail.

Since SIDS seems to involve hitches in a baby's breathing system, the BabySense monitor constantly "watches" the baby's breathing and her every movement. If the monitor senses a pause of 20 seconds in breathing or less than 10 "micro-movements" in the space of a minute, it sounds an alarm, affording caregivers precious time to come and revive the baby.

"In most cases, all it takes is a light intervention to resuscitate a baby," Yaniv says. "But in all cases, time counts, and the BabySense is designed to provide those critical extra seconds that could make the difference."

The BabySense monitor has pretty much rendered the old school monitors obsolete, and its effectiveness has made it the state of the art around the world. "I can say this because it was my dad who invented it — it was really revolutionary," Yaniv says. Since its introduction, the company has sold more than 600,000 monitors.

Of course, nothing replaces the vigilance of adults, but as anyone with kids will tell you, parents and caregivers can use all the help they can get. In 32 countries around the world — from the US to Australia — the BabySense monitor is providing that help.

As the BabySense monitor enters its third decade, it's not possible to know exactly how many children it has saved, nor how many parents' minds it has put at ease. But until the day when we nail down a cause for SIDS and a protocol for preventing it, the BabySense monitor will continue to keep babies sleeping more safely, and their parents more soundly.

13. Drip Irrigation: Water the Desert

In the words of my generation, let's get real.

There are now more than 7 billion people coating the surface of Planet Earth, and that number's going to go up a lot more before it goes down. The rankings of what people consider to be life's necessities will vary depending on whether they live in Greenwich or Ghana, but the one thing that's at the top of everyone's list is water.

Earth is 2/3 water, but only 10% of that is the fresh water we need for drinking, running machinery, growing rice and beans, and shaving and brewing beer and making blue jeans.

Analysts from the UN to the Council on Foreign Relations are painting a fairly grim picture of our ability to keep up with the demand for water in the years to come. The situation is dire — already more than a billion people in Africa and Asia don't have access to clean drinking water — and the challenge is huge . . . but it's not insurmountable.

I suppose that any steps individuals can take to cut down on our use of water would help — shorter showers, less car washing, fewer ice cubes in our mojitos. But when it comes to making a real dent in humans' use of fresh water, if you're not talking farming, you might as well not be talking.

Here's how we humans divvy up the planet's fresh water: Think of ten glasses of water lined up on a kitchen counter. They represent the world's supply of fresh water.

You drink one glass: that's the portion people put to domestic use.

Then a factory owner comes in to the kitchen and grabs two glasses: that's the slice industry uses to do its thing, to cool engines and make paper and fabric and so on.

Then a farmer comes in, grabs the remaining seven glasses, takes them back to his farm and pours them on his crops.

That's right. Seventy percent of the available fresh water on our planet goes to agriculture.

And the vast majority of that water is used not only unproductively, but in ways that do more harm than good.

This is the challenge that the legendary Israeli firm Netafim has been confronting since the 1960s — that challenge of bringing agricultural irrigation in line with a drying planet.

"Let's be clear," says Naty Barak, a director at Netafim, "most agriculture in the world is not irrigated at all. That's really poor resource management. The bulk of the irrigation that is done is incredibly inefficient."

Irrigation breaks down into three main categories: flood irrigation, sprinkler irrigation and the method that Netafim pioneered, drip irrigation.

Flood irrigation (essentially pouring water on crops) accounts for almost 80% of the irrigation that's done on farms, and according to Mr. Barak, it's harmful to soil (the water pushes the soil all over the place) and water sources, produces diminished yields, and is wildly wasteful.

The sprinkler family of irrigation devices — used for about 15% of farm irrigation — is a bit better than flood irrigation in terms of efficiency, but that's not saying much.

Bringing up the rear in terms of usage is drip irrigation, which is the method of choice for about 5% of farms. But it is the unquestionable state of the art in terms of efficiency and its ability to squeeze the most productivity out of every drop of water.

Estimates of the water savings drip irrigation provides vary, largely because watering conditions vary, but they come down between twenty and fifty percent over sprinklers, the next most efficient method. In other words, when it's really working right, drip irrigation can effectively double your water supply.

Drip irrigation might not sound like your idea of scintillating cocktail party chitchat, but the story of its discovery is a piece of Israeli lore.

Israel isn't the #1 driest country on earth, but neither is it Seattle. So Israel has more than a bit of history with irrigation. And from the beginning, it was always a matter of moving water around and then turning on the spigot.

In the 1930s, a water engineer by the name of Simcha Blass was visiting a friend in the desert when he noticed a line of trees with one member that was noticeably taller and more robust looking than the others. He did a little digging, literally, and noticed that a household water line running along the tree line had sprung a small leak in the area of that one tree and was feeding it with a steady drip drip drip of water. The wet spot on the surface didn't seem like much, but down below was a large onion-shaped area of juicy soil.

The idea of drip irrigation was born.

Mr. Blass partnered with Kibbutz Hatzerim in the Negev desert to develop entire drip irrigation systems. He tinkered with variations on the idea, but when plastics became widely available in the 1960s, he finally had the ability to put drops of water precisely where he wanted, when he wanted; Mr. Blass and the kibbutz founded Netafim.

Since then, Netafim has sold its systems in more than 100 countries worldwide. And, according to Mr. Barak, the more we ask of our planet's limited water supply, the more Netafim's systems will benefit the world.

"Water has been declared to be a basic human right," he says, "but we squander it with wasteful irrigation. Drip irrigation provides the ability to make water work harder and more productively than it's ever done in the past."

Mr. Barak makes the point that if 15% of farms using conventional irrigation switched to drip irrigation, the supply of water available for domestic use would double.

That statistic alone should make the world take notice. But Netafim's drip irrigation systems do a lot more than just move water around the farm.

"We're not just talking about a hose with a hole," says Mr. Barak. "The system is very sophisticated, because you need to make sure that the plants that are far from the valve get the same amount as the plants that are close to the valve. You have to be able to maintain consistency in a system that runs uphill, that runs downhill . . . Inconsistency equals waste, and the goal is to eliminate waste."

Mr. Barak adds that the very idea of water has changed since the company started. "In the 1960s, we were using plain drinking water. Today we use recycled water, waste water, brackish water . . . and we're adding nutrients mixed in with the water, so that, in a way, what we're really doing isn't irrigation, it's 'fertigation.'"

Dirt, nutrients, waste — all of this business going on in the water is what makes controlling it so tricky. And this is where the ingenuity of Netafim comes in — in the valves that control the actual drops of water. The valves are spaced precisely along the irrigation lines, and they must work in concert with one another.

"These valves are what make the system work," says Mr. Barak. "They are anti-clogging, self-cleaning, very sophisticated little mechanisms. They make it possible to get greater crop yields, greater crop control, while using significantly less water than we did just a few decades ago.

How much less? In 1965 a typical drip irrigation system could use anywhere from two to four liters of water per hour, which was a vast improvement from the prevailing flood irrigation system.

But now, a typical Netafim system will use half a liter per hour . . . and Netafim is still trying to get that number down.

The main reason, Mr. Barak explains, is simple. "In the next 100 years we are going to have to produce more crops than we ever have, with far less environmental damage than we're doing now."

Some are well on the way to achieving that goal. About 75 percent of Israeli farming is done with drip irrigation, with practically no flood irrigation at all. Drip irrigation accounts for about half of irrigation in California; South Africa is also a big user.

But other areas of the world have yet to make the shift — and food is only one of the reasons why it's important for them to do so. Because when you really dig in to the nuances of drip irrigation, you start to see how widespread its ramifications are.

For example, by using water more efficiently, drip irrigation means you use less fertilizer. Fertilizer production is a significant source of greenhouse gas emissions.

Likewise, the growing market for biofuels will benefit from drip irrigation through the reduced cost of raising fuel crops, which will in turn aid in the spread of biofuels, which will reduce the production of greenhouse gasses, which will, in the long run, reduce the pressure on the world's water supplies.

There's also a quality of life issue at stake here — and I'm not talking about watering lawns. There are places where the challenge is not managing water, but simply getting it. Netafim is in some of these places, and the changes they have made are remarkable.

Mr. Barak tells of one Kenyan village where the women spent the bulk of their day carrying water from a small lake to the fields. With a drip irrigation system in place, not only are they realizing far higher crop yields, but the women are now freed up to spend their time far more productively than they were before . . . getting an education, for instance.

It's an open question whether we can push our existing water supplies to provide us with the food — let alone the ice cubes — to which we've become accustomed. But Netafim has already demonstrated an ability to take a productive technology and keep pushing it well beyond its initial boundaries. There's no reason to think they can't do the same for the world's existing water supply.

14. Luggage Finder

Hit an Amateur Night at any comedy club anywhere in the US and it's a near certainty that at least one of the participants will go off on a tear about the trials of traveling by air. That air travel has become a joke is true to the point of cliché. Let's discuss it for a moment anyway.

It's pointless trying to decide which aspect of the whole ordeal — the lines, the attitude, the food they no longer serve or, worse, the food they do serve — is the most demeaning. Why choose? They're all pretty terrible.

But it feels good to vent, so: for my money, the most infuriating phase of air travel is the dreaded Baggage Carousel Cattle Call. I think because at that point, you've done it. You've completed the trip! You're on the ground! You can leave! But no — actually you can't leave . . . because your bags aren't there.

So what do we do? We circle up and wait around the baggage carousel — a term that makes this assignment sound a lot more fun than it is — and nod heavy-lidded and slack-jawed at each piece of luggage that trundles by, not knowing when or even if our own belongings will at last come around the bend. Escape is so close — the taxis, the bus, your ride. You can see them all through the glass doors. But we have to just stand there and wait, tired, humiliated and, as far as the fate of our luggage goes, utterly ignorant.

According to the FAA, 713 million passengers flew in the US in 2010 . . . and standing there in baggage claim, there are times when

I feel like every one of those passengers is standing there with me, looking for a bag that looks exactly like mine.

It's the last little jab from an industry that increasingly appears to be conducting an experiment to see just how much of their dignity its clients are willing to surrender.

But now, finally, one Israeli company is striking a blow for the common traveler!

The Easy-To-Pick Wireless Luggage Locator — the brainchild of a former Israeli Army general who had spent too much time as a spectator at the baggage claim's suitcase derby — promises nothing less than to unshackle travelers from the captivity of the carousel. The Wireless Luggage Locator is a nifty little device that lets you know when your luggage has arrived and is ready to be picked up — freeing you to do pretty much whatever else you want until then.

"It's one of the worst things about flying," says Yosi Naftali, President of Naftali, Inc., one of Easy-To-Pick's investors and the Luggage Locators worldwide distributor. "That frustration of Is it here, Is it here, Is it here . . . It's much better to be able to go for a walk, do some errands, and relax. Older people can sit down. Travelers with children can get organized. Business travelers can send an email . . . and when their luggage arrives, the locator lets them know."

The Wireless Luggage Locator comes with two components.

The first part is a tag the traveler attaches to his bag, much like an ID tag . . . and here's where this seemingly humble gadget shows its smarts. The traveler switches the tag on when he checks his bag — and the tag senses when it's been loaded into an airplane. It switches off automatically, then reactivates once the bag has been taken off the plane.

The other part is a receiver that the passenger keeps with him.

Once he arrives inside the terminal, he activates the receiver — and it sends him a signal when the bag is within about 100–120 feet of the receiver. The passenger gets the signal and knows that it's time to go pick up the bags.

Simple. Elegant. Vastly better than the guess-and-hope method so many people use right now.

The Wireless Luggage Locator is in use, right now, at airports around the world, and travel-centric international companies like American Express, Ltd., have adopted it as a pointedly representative promotional gewgaw.

But despite its simplicity, the device did present challenges. Chief among them, says Mr. Naftali, were getting the code pairings right so that only one tag would work with one receiver — and do so among all the other competing devices in use at a busy airport.

But now that they've got that problem solved, the company is looking at the next step in the device's evolution. Next up, Mr. Naftali says, Easy-To-Pick is looking to build transmitters directly into luggage; the company is in negotiations with several manufacturers.

It won't make your flight shorter, it won't make your seatmate less chatty, but the Wireless Luggage Locator will make flying a little easier, a little less nerve-racking, and a little more graceful. And in today's travel environment, anything that can do that ought to have an airport named after it.

15. Saving the Bees

There's a bumper sticker going around that says "Go confidently in the direction of your dreams. Live the life you've imagined." The bumper sticker tends to attribute this quote to Henry David Thoreau. While I don't disagree with the sentiment, I'm not convinced that Thoreau actually wrote it, at least not in so many words. (He does express this idea in Walden, but not in such modern, self-helpy terms).

Another quote has made the rounds in recent years that has none other than go-to genius Albert Einstein saying "If the bee disappears from the surface of the earth, man would have no more than four years to live." Again, I'm a little dubious as to whether Einstein — a physicist, not an entomologist — ever said such a thing.

However, even if it takes some license with its author, I'm happy this little aphorism is getting out there because of the idea it introduces. The situation might not be as dire as complete human extinction between Olympics, but there is no question that without bees, it would be much harder for humans to feed themselves.

Eyal Ben-Chanoch *really* doesn't want that to happen. "About one third of the food people eat is dependent on bees," says Mr. Ben-Chanoch. "And bees are in trouble."

The problem is that bees are disappearing, not just individually but by the hive — on the order of 30% of professionally maintained hives per year. The condition is called Colony Collapse Disorder (CCD), it's happening all around the world, and it's ominous.

Mr. Ben-Chanoch is CEO of Rehovot-based Beeologics, Inc., that positions itself as the guardian of worldwide bee health. Beeologics' primary mission is to do all it can to eradicate CCD and restore global bee populations to robust health, all by applying brand new thinking to the way bees are handled.

Mr. Ben-Chanoch says that it's time we re-think bees. "Bees are an animal that make food," he says, "but we don't treat them like that. We take them for granted. But if something was wiping out every third cow, believe me there would be an uproar."

Mr. Ben-Chanoch's primary motivation in helping bees is to support another population, one that he is even more worried about — beekeepers.

"Bees are not going to go extinct," says Mr. Ben-Chanoch, providing a bit of cold comfort. "Eradicating entire species of insects is difficult primarily because there are so many of them. But CCD means the economics of maintaining hives will make the business increasingly unsustainable. So more and more crops will be dependent on fewer and fewer hives. Eventually, if beekeepers keep losing thirty percent of their bees every year, they won't be able to stay in business . . . and then people all over the world will suffer from severe food shortages."

Those are the stakes. To see how they play out, take the case of almonds. The US supplies eighty percent of the world's almonds, and almost all of those are grown in California. The 2010 crop was valued at $2.7 billion — real money by anyone's standards.

While some plants can self-pollinate or take advantage of external conditions like wind to pollinate, almond trees are completely dependent on pollinators to fertilize them, which is the process that coaxes forth the almond nut. While everything from butterflies to crickets can act as a pollinator, for the most part it means bees. Without bees, there's pretty much no almond crop.

Since there aren't enough bees naturally occurring in California's central valley — the valley of the almonds — to keep the trees pollinated, the practice has been for professional beekeepers to truck in their bees and park them in the fields for a couple of months. This lets the bees get acclimated and get out into the fields to do their work.

Annually, about 1.4 million hives are sent to California — about sixty percent of America's managed beehives. When you consider that number, the problem becomes stark. Placing that many hives in one spot for upwards of 10 weeks at a time means they're not available to pollinate in the rest of the country. And when beekeepers are staring down such a significant loss of their "livestock" every year, they start to look at different ways of making a living. Without beekeepers, life becomes very difficult for farmers. And then it becomes very difficult for the rest of us.

Looked at in this way, bees start to resemble a vital natural resource — and we're losing them at an unacceptable rate.

Completely solving CCD is no small order, partly because there appears to be no single cause for CCD. "The jury is out on the causes of CCD, and I think it will always be out," says Mr. Ben-Chanoch. "Bees are subject to all kinds of pressures. But we think that the main line of attack is viral, and that's where Beeologics has developed solutions."

While Mr. Ben-Chanoch is right about the difficulty of pinning CCD to a single cause, he's also right in saying that viruses seem to be one of the primary drivers of colony collapse. And so the biotechnologists and virologists at Beeologics have developed a line of products designed to "immunize" bees from their main sources of infection, including the Israeli Acute Paralysis Virus (IAPV).

Beeologics' Remembee is the most fundamental of these

products; the others are more specialized. Beeologics developed it by applying genetic technology at the cellular level. It's all highly technical — in fact it builds on research that won the 2006 Nobel Prize in Physiology. But the main idea is that when you administer Remembee to bees, usually as an additive to the supplemental syrup solution all managed bees are fed by their keepers — you "immunize" them against the IAPV and other viruses that have been traced to CCD.

So far, the results have been encouraging. "We have immunized thousands of hives around the US over the last three years, and the results are incredibly encouraging — so far," says Mr. Ben-Chanoch.

Mr. Ben-Chanoch cautions that Beeologics is still in the early stages of its work, and that measuring the effectiveness of its products is made trickier because of the nature of what Beeologics is trying to achieve. "You're trying to prove a negative," he says, "that what you did kept something from happening. This is tough. And there are so many other variables. One cold snap can wipe out some colonies and skew your numbers. But so far, we are getting fantastic results."

Mr. Ben-Chanoch can cite survival statistics and correlate them to environmental conditions, but one ground-level finding strikes me as the most telling: Beekeepers who have gone through the first year's supply of Beeologics products have been the quickest to sign up for Beeologics' next round.

Further proof of the concept: Beeologics was recently purchased by the Monsanto Corporation, which probably wouldn't have placed that bet if they didn't think the company was onto something.

For now, Beeologics is confining most of its work to the US, primarily because the regulatory hurdles here are so high. But, "the rest of the world is watching," says Mr. Ben-Chanoch. "If we can make a difference here — and things seem to be going that way — then the rest of the world will say 'okay, us too.'"

Whether Thoreau or Einstein realized it, people are going to go in the direction of the bees. Thanks to the work of Beeologics, it looks like we can do so confidently.

16. Breathing Like a Fish

We often think of inventors as nutty old timers working madly in the barn until something explodes. But adults aren't the only ones who look at the world and see room for improvement. A lot of inventions — from the Popsicle and earmuffs to a basketball that shows you how to shoot it — were the brainchildren of actual children. And I'd bet that for every contraption actually cooked up by a child, there are hundreds that were inspired by children who posed a challenge to a resourceful adult.

One such case: Like-A-Fish, a truly cool underwater breathing apparatus developed by Israel's Alon Bodner.

According to Mr. Bodner, he and his seven-year-old son were watching a Star Wars movie in which the heroes executed a deep water dive using only a tiny little mouthpiece. Mr. Bodner's son asked him if what they'd just seen was possible.

Now most parents would respond with something from the shopworn "it's just a movie" line of parental come-backs. But Mr. Bodner happens to be an engineer and a SCUBA diver, so rather than dismissing his son's question, he thought a moment and said, "I'll get back to you."

The puzzler from his son prompted Mr. Bodner to look at the problem of humans in the water from the perspective of someone other than a human — namely, a fish.

Fish, Mr. Bodner explains, are air-breathing creatures just like us. The only difference is they breathe air that's dissolved in water. "Water naturally contains dissolved air, including oxygen. There's plenty of air in the water for fish to breathe," he says. "So there's air for us to breathe as well."

When we humans go for a dive, we take along cylinders of compressed air, which works well, except that we can only stay down as long as the air lasts. Mr. Bodner's system will enable divers to stay down virtually indefinitely. This has major implications not only for recreational diving, but also for industry, telecommunications, and — since we're already in the realm of sci-fi — underwater dwellings.

Like-A-Fish — which currently exists as several proto-types — operates by extracting dissolved air from the water, and transforms it into the gas form that we can breathe. While this sounds (to me at least) like a miraculous accomplishment, Mr. Bodner says that the main obstacle to developing the system was not the mechanics and the chemistry of the process, but rather its energy consumption.

"There are two main ways to derive a breathable gas from water," says Mr. Bodner. "One is by separating water molecules into oxygen and hydrogen. This requires an enormous amount of energy — an amount that only a nuclear reactor can provide." In fact, this method is how nuclear submarines produce air for crews to breathe.

For everyone else, it's a question of either bringing air along for the ride, or digging it up out of the water itself.

"Technically, it's a very interesting project," Mr. Bodner says, "as it includes basic chemistry and physics, bubble theory, energy calculations, diving physiology and mechanical design — for starters."

The heart of the Like-A-Fish system is a centrifuge that spins a container of water, creating areas of low pressure where dissolved air escapes, is captured, and is converted into a breathable gas.

It's a brainful to grasp for a lay person like me, but for a useful example of how dissolved gas escapes from a liquid when the pressure on the liquid is reduced, think of what happens when you unscrew the cap on a soda bottle. That PPFFFFFFFT sound is dissolved gas (C02) leaving the soda and entering the atmosphere.

As Mr. Bodner refines his system, he has been fielding inquiries from the Israeli and US navies and virtually every manufacturer of diving equipment. And while his initial goal was to create a system that would free divers from the burden of SCUBA tanks, Mr. Bodner has discovered that the juicier market might lie in larger-scale users.

"Conventional systems are open-water systems," he explains. "This means you inhale from a tank and exhale into the open water. This requires a very large quantity of air, which makes it impractical for setting up a dwelling under water. But Like-A-Fish can be used as a closed system, which recirculates the air, as on a submarine — only with a theoretically inexhaustible supply of air, so it's perfectly suited for life under the sea."

Mr. Bodner estimates that there are 1500 Diesel submarines and small submersibles at work in waters around the world, and they're severely restricted in bottom time because they use conventional air supplies.

And, he says, many academic institutions see big possibilities for Mr. Bodner's technology in their long-term research facilities. Translation: those underwater dwellings. There are about 60 in operation around the world, and they are dependent on surface ships to either pump air into the facility or swap out their air bottles. Mr. Bodner's system can relieve them of that burden.

I don't have statistics on this, but I'd guess that when kids are asked to draw a fantasy house, a meaty percentage of them give it a rocket launcher and a wall-sized TV, and place it under water.

It might not be time to hire a deep sea moving van just yet, but as with previous outlandish ideas that went from make-believe to here-and-now (like flip-phones and the Internet), and thanks to Alon Bodner, the time when people can easily move around in our planet's oceans, and even make a home there, has definitely moved closer to becoming a reality.

17. Freezing Breast Tumors

It's indisputable that the increase in the global awareness of breast cancer has been a good thing. Here in America alone, breast cancer has its own ribbon, its own color (pink), even its own month (October). According to breastcancer.org, death rates from breast cancer have been decreasing since 1990 — especially in women under 50. A lot of that can be chalked up to treatment improvements and early detection, all of which can be traced at least in part to increased awareness.

So consider this: every time a woman performs a self-examination or gets a screening from a healthcare professional, there's a chance that the search will turn up a lump. But not all lumps are cancerous or malignant. In fact, according to *US News and World Report*, 80% of lumps are benign — this in spite of the fact that on the order of 300,000 American women undergo a surgical breast biopsy every year. That means 240,000 women per year are undergoing surgery unnecessarily, because once a lump is determined to be benign, doctors will just leave it where it is.

Now granted, finding out that you don't have cancer is far better than finding out that you do. However, once a lump is determined to be benign. . . . then what? Women still have to undergo breast exams, and any time they feel a lump, they're left to wonder: "is that the lump I already know I have? Or is it something new — and something to be worried about?"

And so . . . the increase in breast cancer awareness has had the unintended consequence of also ratcheting up the anxiety level of countless women around the country and the world.

Logically, the best way to eliminate this guessing game would be to remove benign tumors, also known as fibroadenomas, from a woman's breast. But until very recently, benign tumor removal involved serious surgery and carried more risks than rewards, including scarring, the necessity of anesthesia, and the potential for breast deformities.

But those concerns are now largely a thing of the past thanks to a Caesarea-based company called IceCure Medical and its new device, the IceSense3, which provides doctors with the ability to eliminate benign breast tumors through a procedure known as cryoablation.

IceCure Medical CEO Hezi Himelfarb says that the difference between treating fibroadenomas surgically versus using the IceSense3 is like comparing a donkey cart to a rocket.

"Our new cryoablation treatment is an office procedure, done with local anesthesia," says Mr. Himelfarb. "It results in no scar, no pain, no breast deformity, no down time, and it typically takes five to ten minutes. After a patient gets up from the bed she can go right back to work, to the gym, home, a restaurant, wherever. And now she knows that the lump that had been worrying her is gone."

In its simplest form cryoablation means destroying tissue by freezing it. "Cryoablation uses very precise instruments to freeze tissue to the point where it is now dead tissue," Mr. Himelfarb explains. "And once it's dead, our bodies know how to get rid of it. Depending on the size of the tumor, the body can absorb a cryoablated fibroadenoma in days or, in rare cases, a couple of months at the most."

While IceCure Medical did not invent cryoablation — the technology has been around in its basic form since the 1980s — the company re-thought what Mr. Himelfarb calls the treatment "platform" to make it far more compatible with the outpatient environment that patients want today.

"Having a benign tumor is better than having a cancerous tumor," Mr. Himelfarb asserts. "But it's better not to have any tumor at all. So we looked at the stumbling blocks that kept women from having these tumors removed. And the existing cryoablation units were simply too big, too cumbersome, too clunky to work in a smooth and seamless way. They were preventing women from seeking treatment."

In the past, Cryoablation had been achieved by applying high-pressure, super-cooled gas directly into a tumor. IceCure changed the game by switching out the gas for liquid nitrogen, which enabled the company to build a unit that is compact enough to fit into a small treatment room in a doctor's office.

IceCure's use of liquid nitrogen resulted in another major advance. Mr. Himelfarb says, "When a doctor performs a treatment with our system, she uses a probe to penetrate the tumor. When she starts freezing the tumor, the probe produces an iceball that can be visualized on a standard office ultrasound. So it can be used in any office with an ultrasound machine."

By reducing the size of the equipment, allowing for visualization using ultrasound, and using needles that require only local anesthetic versus the general anesthesia demanded by surgery, Mr. Himelfarb says "we have now taken cryoablation out of the operating room and turned it into an office-based treatment."

The IceSense3 system not only presents advances in convenience over previous generations of cryoablation systems. It's also a big improvement over the other major form of non-invasive tumor treatment: heat ablation.

It turns out that tumors are as susceptible to extreme heat as they are to extreme cold. However, heat ablation — usually accomplished by laser, microwave or High Frequency Ultrasound (HIFU) — has a number of drawbacks that make it an unattractive choice for women. For one thing, whereas the cold involved in cryoablation acts as its own painkiller, heat ablation is very painful, meaning it has to be done under general anesthesia.

Also, cryoablation creates an iceball which a doctor can clearly see and manipulate during the procedure. The ice can be targeted very specifically to select for areas that the doctor wants to treat and protect the areas that the doctor wants to leave intact. Heat is much harder to control — our bodies react to it in different ways from person to person and even from tissue to tissue. The only way to see it effectively during a procedure is to use an MRI machine, which is big and cumbersome compared to the IceSense 3.

Finally, heat ablation requires a team of healthcare pros. The IceSense3 treatment can be performed by a single physician or radiologist — anyone trained to operate a biopsy needle.

IceCure Medical has launched the IceSense3 system in and around Cleveland, Ohio, where it has been used in hundreds of procedures. The next few years will see a nationwide rollout of the system. It seems clear that the IceSense3 represents the direction where the treatment of fibroadenomas is heading. So the question follows: what about more insidious cancerous tumors?

According to Mr. Himelfarb, most fibroadenomas are oval shaped. The IceSense3 probe creates an iceball that is roughly oval shaped, so the fit is natural. Cancerous tumors, however, tend to be highly irregular in shape. And, as Mr. Himelfarb says: "Our needle freezes tissue at minus 170 centigrade. Nothing can survive that, so our iceball definitely can kill cancerous tissue. The problem is that once you start to form the iceball, it becomes difficult for a physician to continue to visualize an entire cancerous tumor, because of all the twists and turns they take in the body. With a benign tumor, if you miss something and only kill 99% of it, the remaining one percent isn't a problem at all. But with cancer, if you've left something behind, you haven't done any good."

However, Mr. Himelfarb goes on to explain that cancerous tumors tend not to become irregular in shape until they progress past a certain size — about two centimeters. This means that the IceSense3 can be used in its present iteration to treat tumors below that size — anything down to one or two millimeters. In terms of how small the IceSense3 can go, Mr. Himelfarb says, "If we can see it on an ultrasound, we can destroy it."

Physicians are beginning to see the potential of the IceSense3 system in the treatment of smaller cancerous tumors. Mr. Himelfarb cites the example of Dr. Eisuke Fukuma, a Japanese doctor who is using cryoablation to treat cancerous breast tumors. Following a single meeting with the IceCure team, Dr. Fukuma purchased two IceSense machines and put them into use as part of a clinical study of cryoablation treatment involving 30 Japanese women with tumors up to one centimeter in diameter.

As for tumors over two centimeters, Mr. Himelfarb says that the company is working on new technology that will enable physicians to go after larger tumors that present more immediate challenges to doctors and patients.

Given the strides the company has already made, there's reason for hope for women all along the breast tumor spectrum. Today: a treatment for a common affliction among women of all ages, one that provides enormous peace of mind with almost no discomfort or inconvenience. Tomorrow: the possibility of taking the treatment of cancerous breast tumors to a whole new level of safety, convenience and effectiveness.

It's still too early to celebrate, but with the way things are going, one day we might be able once again to let October be about hay rides and cider, reserve pink for our fuzzy slippers, and do away with the ribbons altogether.

18. Earth-Friendly Food Packaging

Here's the problem with plastic: as a packaging material, it's really really good.

Plastic is waterproof, airtight, and it doesn't let spoilage in. In fact, it's so good at being tough and impervious to the elements that you kind of can't throw it away.

Plastic containers more or less do not biodegrade. When exposed to sunlight over long periods of time, they will break down into their constituent parts — but that's it. They never really "return" to the Earth. Rather, they degrade into harmful chemicals that leach into soil and water and make life messy for anyone who encounters them.

But for the time being, Americans (and the world) have thrown their lot in with plastic. These stats from the Clean Air Council provide a snapshot of our commitment:

- Every year, Americans use about 1 billion plastic shopping bags, creating 300,000 tons of landfill waste.
- Less than 1 percent of plastic bags are recycled each year. Recycling one ton of plastic bags costs $4,000. The recycled product can be sold for $32.
- When the small particles from photodegraded plastic bags get into water systems, they are ingested by filter feeding marine animals. Biotoxins like PCBs that are in the particles are then passed up the food chain, including up to humans.
- In 2008, only 7.1% of disposed plastics were recycled.

- About 31% of Municipal Solid Waste generated in the US in 2008 was containers and packaging, or 76,760 thousand tons. Only 43.7% of that was recycled.
- In 2008, the average amount of waste generated by each person in America per day was 4.5 pounds. 1.1 pounds of that was recycled, and .4 pounds, including yard waste, was sent to composting. In total, 24.3% of waste was recycled, 8.9% was composted, and 66.8% was sent to a landfill or incinerated.
- Newer, lined landfills leak in narrow plumes, making leaks only detectable if they reach landfill monitoring wells. Both old and new landfills are usually located near large bodies of water, making detection of leaks and their cleanup difficult.

So . . . while plastic packaging does an admirable job of keeping our strawberries fresh and our sodas fizzy, it would be an understatement to say that our dependence on it comes with some tradeoffs — mainly involving the health of our water, soil, air and selves. And we're going to be making those tradeoffs for the hundreds of years it takes some plastics to break down and become benign.

Wouldn't it be great if someone came up with food packaging that goes back to nature in a matter of days, rather than centuries?

That was the question that Daphna Nissenbaum, an Israeli computer engineer and mom, asked herself when she realized one day that dealing with food packaging was taking up far more of her time than she was proud of.

"In the course of my day I found myself spending all this time dealing with recycling, taking bottles to the recycling center, searching for somewhere to recycle a plastic bottle, or washing and reusing it when there was nowhere to take a package — my life was becoming consumed with empty packages."

In April of 2010, the reality of her package-intensive lifestyle motivated Mrs. Nissenbaum and her partner Tal Neuman to start Tipa Corporation, a Ramot-Hashavim-based company that creates food packaging materials that are 100% biodegradable.

"Our goal is to create packages that imitate the orange peel," says Mrs. Nissenbaum. "When you eat an orange, you just drop the peel and go. The peel goes back to nature within a few weeks, and the Earth is blissfully oblivious. Our packages are precisely the same. They are completely harmless to the environment, and when left out in the open, they'll be totally gone within a few years at the very most — but more likely within a few months."

Tipa's line of materials can be used for drinks and dry snacks and are lighter and more cost-efficient to manufacture and ship than conventional materials.

Tipa's difference-making, game-changing innovations are highly technical but the upshot is a line of packaging materials that will enable all manner of food packagers to realize up to twelve months of shelf life for their products while producing no down-stream damage to the environment.

Mrs. Nissenbaum said that the key to the company's success took place only partly on the drawing board — Tipa packages rely on a proprietary compound that is layered in such a way as to mimic the preservative qualities of plastic.

The other ingredient of the company's success was stubbornness. Mrs. Nissenbaum explains: "When we first came up with our idea, we were thinking about aesthetic designs that would appeal to kids, to adults, to athletes, to moms, and so on. In the meantime the experts we hired were working on chemical designs. After 6 months, they said 'it can't be done.' We said 'no way, we're doing this', and we hired a new researcher and stayed on it until finally we had our breakthrough."

Tipa (the name is Hebrew for "a water drop") is just starting out in the world, but already the company has been recognized for its breakthrough technology, including taking second prize at the packaging industry's 2012 SusPack (Sustainable Packaging) Awards.

Starting in the first quarter of 2013, Tipa will spend the next two years rolling out its product line in Australia and Europe — but its biggest push will be in the US. Mrs. Nissenbaum says that Tipa has already signed deals with some of the biggest American food companies; they'll begin ramping up production by the middle of 2013.

"The US has the greatest need for our products," says Mrs. Nissenbaum. "The US generates the most waste per person. But fortunately, Americans have a greater awareness of the problem than anywhere else. American consumers are more concerned and more active in the area of recycling than even in Europe."

Still, reducing waste is only part of the equation when measuring the impact of packaging materials on the environment. Producing plastics is an intensively polluting business; the creation of plastic demands an enormous amount of water and energy. To put the issue in perspective, consider that PVC is the third most widely produced form of plastic. According to the Clean Air Council, producing a year's worth of just the chlorine necessary to make PVC uses as much energy as eight medium-sized nuclear power plants.

Mrs. Nissenbaum is acutely aware of the damage food packaging production can do, and she states that Tipa products don't only provide benefits in their afterlife. "Part of our mission was to create packaging materials that are an improvement over existing materials at every stage of their lives, not just after they've been used," she says.

"We're still gathering precise numbers, but overall our carbon footprint is definitely lower than that for traditional plastics, along with other resources used in their production."

Nutritionists will tell you that the closer food is to its natural form, the better it is for you. Environmentalists will say the same thing about the wrapper. Tipa is enabling the world's eaters to have both.

19. Restoring Macular Vision

The universe is big.

I know this because I once looked at it on a cool night in a desert setting when the humidity was low and the sky was cloudless. Also, I was using a telescope. It was this last element, of course, that enabled me to see things I had never seen before — the rings of Saturn, Jupiter's red spot, a nebula or two.

As I marveled at all that was going on "out there," I was struck by the thought that the telescope might be humankind's single most powerful tool for putting our everyday lives into perspective.

Back when I was standing there shivering in the desert, most people — myself included — only thought of the telescope as a device that helps us see beyond the scale of what we experience on a day-to-day basis.

But in 2010, a new kind of telescope was cleared for use in the USA. This telescope is designed to be surgically implanted in the human eyeball. Its goal is to treat people with age-related Macular Degeneration (AMD), a disease that gradually destroys the central vision of the elderly. And it holds the potential to help millions of Americans see what's right in front of them.

AMD affects almost 2 million Americans. Each year something like 500,000 more people are diagnosed with the disorder. It's the leading cause of legal blindness in American adults over the age of 60, and as the US population ages, AMD is only going to grow more prevalent.

Patients with AMD suffer from the accumulation of fatty deposits on the retina — the light sensitive area at the back of the eyeball where (to put it crudely) images are gathered and sent to the brain for processing. Over time, these deposits result in a blurry spot in the center of a patient's visual field. The result is that, were you suffering from AMD while trying to read this book, you might be able to make out the edges of the pages and some of your surroundings, but the center of the page would be obscured by an opaque greyish blob. AMD makes simple things like driving, answering a phone or recognizing faces next to impossible.

There is at present no cure for AMD. There is, however, a treatment based on a truly remarkable new device: the WA2.2X Implantable Miniature Telescope.

The telescope was invented by Dr. Isaac Lipshitz, an Israeli ophthalmologist with more than 30 years of experience treating vision disorders. Dr. Lipshitz describes himself as someone who always felt that the lack of a cure for a disease doesn't have to be the end of the story.

"With many patients with retinal diseases, getting a diagnosis is no help, because there's no cure," he says. "What I wanted to do was not cure the disease but make the best of whatever vision these patients have left."

His insistence on not taking no for an answer resulted in the development of a telescope that can be implanted directly into the eye of a patient suffering from AMD. The telescopic implant is about the size of a pea, it sits squarely in the middle of the eyeball, and once it's in, it can help a patient return almost seamlessly to a normal range of activities.

Though the technology has just become available to Americans, its roots go back decades. Dr. Lipshitz says that in the 1980s many enterprises throughout Israel, the US and Russia were experimenting with "ocular telescopes" that would connect to the brain. "But we couldn't make it work. Computers weren't fast enough to make the calculations we needed, and there were a lot of other problems with external devices."

Dr. Lipshitz's breakthrough was his determination that to make the idea work, he had to develop an entire telescope that could be implanted in the eyeball.

"Of course, everyone thought this was crazy. No one had ever implanted a device in the eyeball. Lenses had been implanted to treat cataracts, but this was a device designed to mimic the natural lens of the eye, and people said, basically: no way."

Dr. Lipshitz proved the naysayers wrong, and in the 1990s, after going through several iterations of the device, he finally had a telescope that worked.

And what it does is really jaw-dropping. AMD causes the central part of the retina — the macula — to become essentially useless. However, the surrounding retina — where peripheral vision takes place — is still perfectly good. So the telescope implant magnifies the image that a normal eyeball would see and projects it onto the peripheral retina.

The result is a useable central image where there wasn't one before. Because the telescope implant is so new in the US, there are some growing pains associated with the treatment.

The surgery itself is fairly straightforward, but Dr. Lipshitz cautions that specialists in both cataract and corneal surgery are the most qualified to perform it. To date, roughly 50 Americans have received the implant, and in the early stages of the implant's availability in the US, roughly 80,000-100,000 AMD patients will be eligible to undergo the procedure.

And once the surgery is complete, patients need to go through a rehab program that teaches them how to use their new vision. Since the eyeball with the implant will be responsible for central vision, it's up to the other eye to cover a patient's peripheral vision. This takes some getting used to. A post-surgical program that Dr.

Lipshitz likens to physical therapy following knee surgery trains patients over the course of about six weeks to use their new vision.

But with a little time and effort, patients can regain an enormous amount of vision — and the independence and self-confidence that goes with it.

But while the telescopic implant is certainly a positive for American AMD sufferers, there is even greater hope just over the horizon — specifically, in Europe, where Dr. Lipshitz has already successfully implanted even smaller, more powerful telescopes in AMD patients.

One such implant is 25 percent of the size of the current US-approved implant. It allows patients to see both centrally and peripherally in the same eye.

Another implant — the Ori (named after Dr. Lipshitz's son) — fits in patients who have had cataract surgery. At 1.25 mm thick, the Ori is so small that it's on course to be named the world's smallest telescope by the Guinness Book of World Records.

Dr. Lipshitz is also working on protocols to make testing patients to see if they're qualified for the implants much easier, replacing lengthy in-office visits with a 5-minute computer test that can be taken, essentially, anywhere.

And currently on Dr. Lipshitz's work table is a prototype for an implant that patients can actually turn on and off themselves, allowing them to alternate between close-up and peripheral vision at the flip of a switch.

Space telescopes can provide jaw-dropping, mind-popping views of almost unimaginable beauty and profundity. I imagine that for AMD victims who have their eyesight rejuvenated by Dr. Lipshitz's telescopes, even if all they're doing is watching a sunset or spending time with their families, they'll find their views no less spectacular.

20. *The* Flash Drive

Raise your hands if you've ever been in Dov Moran's shoes.

In April of 1998, the Israeli entrepreneur was flying from Israel to New York City to make a presentation on behalf of his company, M-Systems, to a convention of Israeli public companies. The flight departed at 1:00 AM, and after settling in, Mr. Moran reviewed his presentation a few times, then turned off his computer and went to sleep. Or so he thought.

"The engineers told me that the computer battery had more charge left than I thought, so it woke the computer back up," recalls Mr. Moran. The result: his computer ran throughout the flight until the battery was completely dead, taking his presentation with it.

It's an experience almost every computer user has had. You push that button, expecting the computer to fire up or the file to burst open on the screen. Instead, nothing happens. There is perhaps no colder feeling in our data-driven world.

For his part, Mr. Moran was able to wing it that day, enough to make a credible presentation. But the experience was a wake-up moment for him. "I was determined never again to come to a lecture without a copy of the presentation in my pocket," he says.

What Mr. Moran wanted was a way to back up his presentation quickly and keep it on hand in a neat, accessible, unobtrusive package. With a background in computer engineering, Mr. Moran knew the existing options for computer data storage were limited. Floppy disks: too small to hold an entire presentation. CD hard disks: too expensive, and too clunky to write on efficiently. External drives lacked elegance and leanness.

But Mr. Moran also knew that, soon, every computer coming to market was going to be using a USB port — a little slot that would enable the computer to connect easily and quickly to an external device.

Mr. Moran was convinced that the USB was the way in to the solution he sought. Mr. Moran started with the idea of a device that could connect to the USB port, and from there he developed the DiskOnKey "flash memory drive," a small device about the size of your thumb that connects to a computer via the USB port and acts as a clean, simple memory and data storage system.

USB flash drives perform the same data storage tasks as old CD-ROMs and floppy disks, but they are much smaller, lighter, more durable, and have thousands of times the capacity of the old formats. Current flash drives can store up to 256 gigabytes of information, with more to come.

To say that the USB flash drive is ubiquitous is an understatement. If you have never used a USB drive, then look to your left and look to your right; someone you just glanced at most certainly has. Mr. Moran estimates that 51 million of them had been sold worldwide by 2004, when the device was just hitting its stride.

Today, flash drives are so commonplace that they've been turned into novelty products. Search the Internet for 60 seconds and you can find USB flash drives embedded in fake dog biscuits and plastic sushi rolls. Companies give engraved flash drives away as free promotional products. They're everywhere.

But for a device that the world was clearly waiting for, getting it off the ground was no day at the beach.

"We put the device on the market in November of 2000 with the backing of IBM," Mr. Moran recalls. "But at that point, IBM was no longer in its glory days. I really wanted to partner with Dell. I made a presentation to them, and they dismissed me because in their minds, the USB flash drive was too new. 'Come back to us when you've proven yourself,' they said."

Mr. Moran says that one year later, he had sold $6 million worth of devices, and Chinese companies were paying him the backhanded compliment of selling imitations.

A return trip to Dell yielded a warmer reception, in the form of an agreement to create an internal committee that would investigate the potential for DiskOnKey.

"After 3 months, the committee returned a report stating that the company and our product had no future," Mr. Moran says. Dell's rationale was that someone wanting a small amount of storage could use a floppy disk, and if they wanted large amounts of storage, they could use an external drive.

"The report was so good, so thorough in its reasoning, that it almost convinced me," Mr. Moran remembers. "But the market never read that report, and sales kept growing."

Much red tape and many lawyers later, Mr. Moran became one of Dell's top suppliers, and he partnered with many other companies as well.

Mr. Moran says his moment of deepest anxiety came when the US company Best Buy alerted him that they were sending him a USB drive that had been returned by a customer for being not just defective, but for literally burning up in the customer's computer. According to Mr. Moran, overheating of the drive was a potential problem that had been taken into account in the design process; there was no way that a DiskOnKey flash drive could overheat from use. Before the defective unit even arrived, Mr. Moran sent his team to the lab to try to recreate the heating problem and look for something they had missed. They tried for days, but they couldn't do it. Mr. Moran recalls lying awake at night, unable to sleep.

"There are millions of DiskOnKey drives in use around the

world," he remembers thinking. "What if they all need to be recalled? It will ruin the company . . . "

Mr. Moran and his team members kept at it, trying to recreate the problem. No matter what, they just couldn't cause a DiskOnKey to overheat with anything remotely like normal use. Mr. Moran was preparing to make a report to Best Buy when finally the defective unit arrived — and it turned out to be a competitor's product. Mr. Moran says he can still taste the relief he and his team felt.

The flash drive continued to thrive, eventually putting the floppy drive out of business altogether and becoming the standard in quick, easy, small and mid-scale computer data storage.

Mr. Moran knows that his invention changed the way people do things — but he also knows that technology moves on.

"When it became clear that computers were going to move away from floppy disk drives," Mr. Moran recollects, "the company that had essentially cornered the market on floppy drives called me to come to Japan to talk about ways we could work together. We talked for hours but it was clear there wasn't anything we could do together. While I was standing outside the company's headquarters with its CEO, just the two of us, he said to me, 'I should hate you, you are putting us out of business.' And I said to him, 'Today it's the floppy disk going out of business, and one day it will be the USB drive. That's the way this business is.'"

In 2006, M-Systems was bought by SanDisk, and Mr. Moran moved on to other challenges.

Mr. Moran is right — eventually the world will move on from USB flash memory. But for the time being, that little device makes life easier and more efficient for millions of people around the world, and helps us all sleep a little better, whether on an overnight flight or at home in our beds.

Oh, and you can put your hands down now.

21. Trauma Saver

Time, as Einstein famously pointed out, is relative. Everyday experience bears this out. A ten minute massage feels like it's over before it got started, but ten minutes spent listening to a political debate can seem like a lifetime.

But in the area of trauma medicine, the thinking about time is unanimous: ten minutes is too long.

Doctors and first responders worldwide agree that in the event of serious trauma — everything from an auto accident to a battlefield wound — emergency workers must do everything they can to get a patient to a hospital within nine minutes of the injury.

The gravity of the situation becomes stark when you consider that, according to the Center for Disease Control, trauma (or "unintended injury") is the number one cause of death in Americans between the ages of 1 and 44.

The point is clear: the sooner a trauma victim can receive treatment, the greater her chances for survival.

In the words of one Pennsylvania trauma doctor quoted in the Pittsburgh Post-Gazette: "Time is medicine."

Dr. Omri Lubovksy learned this in 2001, when the Israeli orthopedic surgeon was taking an advanced trauma life support class as part of his medical training.

"The first thing they stress is, in trauma, you have to get the patient into the hospital as fast as possible. But right away, while taking this training, I noticed a problem," Dr. Lubovsky recalls.

Like all who undergo such training, Dr. Lubovsky was taught that providing medical care to someone who has sustained a serious trauma is done in a certain order according to strict guidelines — and that the first task is always to make sure that the patient has a clear airway.

"Opening up the airway in an unconscious patient requires you to perform a procedure called the jaw thrust maneuver," he says. "It's a way of manipulating the jawbone by pulling it forward to open up the back of the throat so air can get in. It's relatively simple and easy to do. The problem is, to keep the airway open, you can't do anything else. You have to use your hands to keep the airway open."

Dr. Lubovsky also found himself troubled by the fact that many trauma patients also sustain spinal injuries, which means they should be moved as little as possible. To stabilize such patients, trauma teams carry cervical collars as part of their aid kits. But performing the jaw thrust maneuver before the collar is put on can put the spine at risk — and waiting for the collar to be put on first wastes precious seconds.

Dr. Lubovsky found himself wondering why there wasn't a device that could do both jobs at once: stabilizing the spine and opening the patient's airway. Such a device, he realized, would also free the medical worker's hands for other jobs.

"In trauma care, it's rare that the only job to be done is to open an airway and stabilize the spine," he says. "Usually there is bleeding to be stopped, broken bones to be stabilized and so on."

Dr. Lubovsky also points out another of the central challenges of trauma care: the problem of the conscious patient.

"Doctors measure consciousness on a scale of 3 to 15," he explains. "Anything below 8, you have to actually insert an internal airway device into the patient. This is known as intubating. Now, intubating an unconscious patient, that takes time, but just a little. But a conscious patient will need to be intubated in case she lapses into unconsciousness on the way to the hospital, which would cause her airway to collapse."

Dr. Lubosvky goes on to explain that a conscious patient will "fight" an intubation procedure tooth and nail. "The world's greatest doctor will require at least 5 minutes to intubate a conscious patient," he says.

And all the while, that nine minute clock is tick-tick-ticking.

Dr. Lubovsky was certain that a single external device could provide for both spinal stabilization and an open airway, freeing an emergency worker's hands and relieving him of the task of intubation.

But it wasn't until he mentioned the idea to his sister Michal Peleg-Lubovsky three years later that the idea became a reality.

"My sister Michal was studying engineering, and needed to do a final project," Dr. Lubovsky remembers. "I mentioned the idea of this special device to her, and she took it upon herself to design a prototype under the guidance of her instructor."

A year of work produced the very first Lubo collar in 2004. The Lubo collar is just what Dr. Lubovsky envisioned: a cervical stabilization collar that also positions the jaw so that the upper airway stays open. Once he saw the prototype, Dr. Lubovsky was convinced that it wouldn't be long before the Lubo collar was out in the field, saving lives. What followed instead, he says, was a lesson in the difference between being a doctor and being a marketer.

"We worked with Hadassah, an organization that takes inventions originating at the Tel Aviv University to market," he says.

It immediately became clear that, since the Lubo collar represented such a radical shift from the accepted way of doing things, the device would need to undergo a lengthy three-stage clinical trial.

The trials began, and the early stages immediately showed promise. But as more results came slowly trickling in over the next few years, other events also took place. The world economic crash. The Bernie Madoff investment scandal. Turnover at medical care facilities. All contributed to slowing the progress of the Lubo collar.

Finally, in 2010, Dr. Lubosvky received word that the third and final stage of the clinical trials had achieved stunning results. The Lubo collar had been used during surgeries on fully anaesthetized patients, and the trial showed that the Lubo collar performed just as well as intubation in keeping patients' airways open.

At just about this time, Dr. Lubovksy and his sister received all rights to their patent back from Hadassah, clearing the way for them to pursue other avenues for bringing the Lubo collar to market.

The next few months brought the Lubo collar to the attention of the Nahal Oz-based critical care device maker Inovytech, which has agreed to — at last! — bring the Lubo collar to the world.

In the wake of the Lubo collar's wildly successful appearance at the 2012 MEDICA international medical trade show, the company plans to introduce the Lubo collar in Europe in April of 2014, followed shortly thereafter by a worldwide rollout. Consider that American Ambulance Association estimates that nearly 50,000 emergency ambulances operate in the USA alone, and you start to get an idea of the size of the market for and, more important, the lifesaving potential of the Lubo collar.

If the Lubo collar lives up to Dr. Lubovsky's vision and becomes part of the standard trauma kit around the world, the seconds it saves trauma crews could add years to thousands of lives.

Conclusion

I think it's safe to say that no country in the world is more controversial than Israel. And no country would have more reason to hunker down and keep to itself. And yet it hasn't.

And so the question remains: why Israel?

Why is this country, a place so ripe for isolationism, instead a hotbed of inventiveness and innovation and pluck and gumption on a scale that is so far out of proportion to its size? Why does it reach outward instead of turning inward? And why do the positive effects of its efforts so dramatically dwarf those of just about every other country on the planet?

I'm far from the first to raise the question. Panels have been convened, college courses have been added to curricula, entire books have been devoted to the matter ("Startup Nation" did a particularly good job of examining Israel's entrepreneurial spirit). But the more I look into it, the more I don't think there's an answer . . . or at least not a single answer.

If you ask the people I asked, you'll get responses as if they were shot from a blunderbuss.

Some say it's the fact that the country has had to innovate just to make life flourish in a desert. Others say Israelis like to hit the books. Some say that requiring military service as Israel does teaches citizens to respect procedure. Others say that Israel's military teaches citizens to mistrust procedure . . .

And so I withdraw the question, because maybe the answer doesn't matter so much. Maybe the facts simply speak for themselves. A certain, possibly unknowable combination of individual characteristics and national conditions have combined to produce an engine of inventiveness that has changed life for everyone else on the planet for the better.

And all indications are that it will continue to do so. As we speak, Israeli researchers are trying to determine if cinnamon can slow or even prevent Alzheimer's disease; they think they've found a way to kill HIV cells; they're developing an implant to cure blindness in the elderly; they've developed a way to make juice and milk cartons spoil-proof; they're fine-tuning a web app that will prevent credit card fraud; they've developed a device that can help paraplegics walk.

I could go on . . . but possibly more important than the gadgets and technologies, the game-changing insights and era-defining protocols that Israel produces, more important than all of that is the lesson that all of this inventiveness teaches — the notion that this planet is a place where good people work and struggle and persevere in an effort to create, to give, to help, no matter what the circumstances.

Many countries, many individuals do this. Israelis just do it in much higher concentrations. And in so doing, they resoundingly disprove the idea that a people are only one thing, that a nation is only this way or that way.

So it's really up to each of us as individuals: we can accept what we see in the headlines, throw up our hands and say "Ah, that place'll never change."

Or you can look closer and see what's really going on there. You wouldn't be alone.

Warren Buffet and Bill Gates are big fans of the happenings in Israel. Buffet: "Israel has a disproportionate amount of brains and energy." Gates: "It's no exaggeration to say that the kind of innovation going on in Israel is critical to the future of the technology business."

Just about every important global IT company has opened an R&D division in Israel. Israel has the third largest number of companies traded on the global stock market, after the US and China.

And according to the American-Israeli Friendship League, Israel, with its population of 7 million, launched 600 startups in 2010, compared to 700 throughout all of Europe — which has a population of 700 million.

Facts are facts. And even the opinions of the Buffets and Gateses are fact-based. Yet . . . so many people look at Israel and see a confirmation of their bias, whatever that bias might be. But if everyone looked more closely, they'd see a place where people go out of their way not only to survive, but to contribute, to improve lives, to make things better.

• • •

People might not see Israel as a place where life exists beyond a certain set of categories and assumptions — or they might not want to see it that way. And as I've mentioned, Israel is not flawless, and Israelis are not saints.

But as I hope the stories in this book bear out, life in Israel is anything but rigid, inflexible or conformist. Instead, Israel is ambitious and positive, organized around progress, growth, improvement and hope for the future. And so are the people who call it home.

So here's one more opinion, my opinion, but it, too, is an opinion that is fact-based: Israel is good for us. By "us" I mean Americans, and I mean Earthlings. Israel has made us — has made you, and people you know — safer and healthier, and it is making life on this planet more sustainable, more stable, and better.

Israel is a friend to you, whether you know it or not.

Timeline

68 Important Israeli Inventions
1948–2012

1940s   **RUMMIKUB:** Ephraim Hertzano invents smash hit board game Rummikub. Rummikub goes on to become the best-selling game in the USA for the year 1977.

1948   **UZI MACHINE GUN:** Major Uzi Gaf develops the Uzi submachine gun. Gaf builds in numerous mechanical innovations resulting in a shorter, more wieldy automatic than had previously been possible. It is estimated that more than 10,000,000 have been built; the Uzi has seen action in numerous wars and in every country around the world.

1950s   **SUPER CUKE:** Professor Esra Galun's research into hybrid seeds leads to his creation of the world's first commercial hybrid cucumber. Their descendants, and the techniques Prof. Galun pioneered account for the majority of cucumbers cultivated today. Prof. Galun went on to develop early-blooming melons and disease-resistant potatoes. His work continues to inform and influence crop genetics.

1954   **CANCER SCREENER:** Weizmann Institute pioneer Ephraim Frei begins groundbreaking research on the effect of magnetism on human tissue. His work will lead directly to the development of the T-Scan system for the detection of breast cancer, which the US FDA described as a "significant . . . breakthrough."

1955   **EARLY COMPUTER:**   The Weizmann Institute's WEIZAC computer performs its first calculation. With an initial memory of 1,024 words stored on a magnetic drum, it is one of the first large-scale stored program computers in the world. IN 2006, the Institute of Electrical and Electronics Engineers recognizes WEIZAC as a milestone achievement in the fields of computers and electrical engineering.

1955   **SOLAR ENERGY BENCHMARK:**   Harry Zvi Tabor develops a new solar energy system that today powers 95% of Israeli solar water heaters and is the standard for solar water heating around the world.

1956   **AMNIOCENTESIS:**   Weizmann Professor Leo Sachs becomes the first to examine cells drawn from amniotic fluid to diagnose potential genetic abnormalities or pre-natal infections in developing fetuses. His work later becomes known as Amniocentesis, a routine procedure now conducted on pregnant women worldwide.

1963   **LAB-BRED BLOOD CELLS:**   Professor Leo Sachs becomes the first researcher to grow normal human blood cells in a laboratory dish. This breakthrough leads to the development of a therapy that increases the production of crucial white blood cells in cancer patients undergoing chemotherapy.

1965   **DRIP IRRIGATION:**   Founding of Netafim, developer and distributor of modern drip irrigation (see Chapter 13).

1966   **COLOR HOLOGRAM:**   Professor Asher Friesem produces the world's first color hologram. He goes on to explore

3D imaging through work that leads to the development of "heads up" displays for pilots, doctors and other virtual reality systems.

1967   **DESALINATION:** Professor Sydney Loeb takes a position at Ben-Gurion University, where he will develop the reverse osmosis desalination process, now the worldwide standard (see Chapter 4).

1970   **ADVANCED CELLULAR RESEARCH:** Ada Yonath, Ph.D., establishes the only protein crystallography laboratory in Israel. She begins a course of research on the structure and function of the ribosome, the sub-cellular component that produces protein, which in turn controls all chemistry within organisms. Her work lays a foundation for the emergence of so-called "rational drug design," which produces treatments for several types of leukemia, glaucoma, and HIV as well as antipsychotic and antidepressant drugs. Along with two colleagues, Prof. Yonath is awarded the 2009 Nobel Prize in Chemistry.

1972   **BLOOD DETOXIFICATION:** Prof. Meir Wilchek demonstrates that "affinity chromatography" — a method he developed for separating biological or biochemical materials — can be used to de-toxify human blood. This work leads to the development of present-day technologies, employed around the world, that are used to remove poison from a patient's blood.

1973   **DRONE AIRCRAFT:** Israeli fighter jets sustain serious damage during the Yom Kippur War. In response, Israel initiates the development of the first modern Unmanned Aerial Vehicles — also known as UAVs or drones. The new

Israeli drones are lighter, smaller, and cheaper than any of their predecessors, with capacities such as real-time 360-degree video imaging, radar decoy capability and increased operating ceilings. Drones enable Israel to eliminate Syria's air defenses at the start of the 1982 war with Lebanon without losing a single pilot. Today drones descending from Israeli designs conduct military, civilian, research and surveillance operations around the world.

1974 **COMPUTER PROCESSORS:** Computer heavyweight Intel sets up an R&D shop in Israel, leading to the development of the globally ubiquitous 8088 processor and Centrino chip.

1977 **RSA ENCRYPTION METHOD:** Professor Adi Shamir, working with two US colleagues, describes a method of encryption now known as RSA. RSA is the single most important encryption method used worldwide to secure transactions between customers and banks, credit card companies and Internet merchants.

1977 **DIGITAL AGE INFORMATION SHARING:** Abraham Lempel and Jacob Ziv develop the LZ data compression algorithms. Aside from their trailblazing academic applications, these algorithms become the primary basis of early computer information sharing. Today LZ algorithms and their derivatives make possible our ability to send many types of photos and images between computers quickly and easily.

1980s **FARM-SCALE FOOD STORAGE:** Professor Shlomo Navarro invents a simple yet paradigm-shifting food storage system intended to help farmers in developing, food-poor and resource-poor areas keep their crops from spoiling after harvest. The system evolves into GrainPro Cocoons, water-

and air-tight containers used around the world to prevent the damaging effects of spoilage and parasites without the use of pesticides.

1981   LEUKEMIA TREATMENT: Professor Elli Canaani joins the Weizmann Institute. His research into the molecular processes leading to Chronic Myelogenous Leukemia (CML) will result in the development of Gleevec, a drug now provided to CML patients around the world. The molecular processes discovered by Professor Canaani were subsequently discovered to be at work in other leukemias, as well as certain tumors and lymphomas.

1981   UNDERSTANDING CELLULAR ACTIVITY: Avram Hershko and Aaron Ciechanover — along with American counterpart Irwin Rose — begin work that will lead to the discovery of ubiquitin, a molecular "label" that governs the destruction of protein in cells. This discovery produces a dramatic improvement in the understanding of cellular function, and the processes that bring about ailments such as cervical cancer and cystic fibrosis. In recognition of their work, the team receives the 2004 Nobel Prize in Chemistry.

1982   A NEW FORM OF MATTER: Israeli scientist Daniel Shechtman discovers Quasicrystals, a "new" form of matter that had previously been considered not only non-existent but impossible. He becomes the object of disdain and ridicule, but his discovery is eventually vindicated and earns him the 2011 Nobel Prize in Chemistry. Applications of Quasicrystals range from the mundane (non-stick cookware) to the arcane (superconductive and superinsulative industrial materials).

1986   **COMPUTER "LANGUAGE":** Computer scientist David Harel develops Statecharts, a revolutionary computer language used to describe and design complex systems. Statecharts are used worldwide in areas from aviation to chemistry. Harel's work is also being applied to the analysis of the genetic structures of living creatures with hopes of applying subsequent discoveries to the analysis and treatment of disease, infection and other biological processes.

1991   **IMMUNOLOGY ADVANCEMENT:** The Weizmann Institute's Prof. Yair Reisner announces the creation of mice with fully functioning human immune systems. Described from an immunological perspective as "humans with fur," the mice provide for the first time a real-world arena in which to study human ailments and represent a major step forward in the search for a cure for AIDS, Hepatitis A and B and other infectious diseases.

1991   **BABY MONITOR:** Haim Shtalryd develops the BabySense crib monitor, which becomes standard child safety equipment in millions of homes worldwide (see Chapter 12).

1993   **OFFICE PRINTER:** Rehovot-based Indigo, Inc., introduces the E-Print 1000. The device enables small operators to produce printing-press quality documents directly from a computer file, revolutionizing the operations of work environments of all stripes.

1993   **COMPUTER SECURITY:** 25-year-old Gil Shwed and two partners establish computer security firm Check Point. Within two years Check Point signs provider agreements with HP and Sun Microsystems. The company experiences phenomenal growth, and in 1996 it becomes the leading provider of firewall and security services — including anti-virus, anti-spam and anti-data-loss security components — to businesses of all sizes around the globe.

1996   **MULTIPLE SCLEROSIS TREATMENT:** Teva Pharmaceuticals introduces Copaxone — the only non-interferon multiple sclerosis treatment. Copaxone — the world's #1-selling MS treatment — helps reduce relapses and may moderate the disease's degenerative progression.

1996   **INSTANT MESSAGING:** Mirabilis launches ICQ, the first Internet-wide instant messaging system. America Online adopts the technology and popularizes the world of online chat.

1997   **COMPUTER DICTIONARY:** Introduction of the Babylon computer dictionary and translation program. Within three years the system will boast more than 4 million users. Eventually Babylon becomes integrated into most user-level Microsoft programs, allowing for seamless cross-language translation of millions of words at the click of a mouse.

1997   **"PORTABLE" SLEEP LAB:** Itamar Medical Ltd is founded, and soon brings to market its WatchPAT sleep lab, representing a paradigm shift in the treatment of sleep disorders (see Chapter 7).

1998   **PILLCAM:** Given Imaging develops the PillCam, now the global standard for imaging of the small bowel (see Chapter 10).

1998   FIRST AID: Bernard Bar-Natan makes the first sale of his Emergency Bandage. A giant leap forward in field dressings, it has become standard equipment in both civilian and military first aid kits worldwide.

1998   NANOWIRE: Researchers Uri Sivan, Erez Braun and Yoav Eichen report that they have used DNA to induce silver particles to assemble themselves into a "nanowire" — a metallic strand 1,000 times thinner than a human hair. In addition to staking out new ground on the frontier of electrical component miniaturization, the wire actually conducts electricity, marking the first time a self-assembling component has been made to function, and laying a path to exponential advances in the field of nanotechnology.

1999   MOBILEYE CAR SAFETY SYSTEMS: Amnan Shashua and Ziv Aviram found MobilEye, a company that provides advanced optical systems to car manufacturers to increase safety and reduce traffic accidents (see Chapter 8).

2000   FLASH DRIVE: M-Systems introduces the flash drive in the US. Smaller, faster and more reliable than floppy disks or CD-ROMs, they will go on to entirely replace those technologies worldwide (see Chapter 20).

2001   ADVANCED UNDERWATER BREATHING TECHNOLOGY: Alon Bodner founds Like-A-Fish, a manufacturer of revolutionary underwater breathing apparatuses that extract oxygen from water (see Chapter 16).

2001   SPINAL SURGERY ROBOTS: Mazor Robotics is founded and goes on to introduce its SpineAssist robotic surgical assistant, the most advanced spine surgery robot in use today (see Chapter 5).

2002   URBAN AIR COMBAT/RESCUE: Rafi Yoeli develops the initial concept for the AirMule urban carrier, combat and rescue vehicle (see Chapter 6).

2002   TERRORIST DETECTOR: In the wake of renewed terrorist activity against Israel and the US, Ehud Givon assembles a team of researchers to develop an advanced and foolproof "terrorist detector," resulting in the WeCU security system (see Chapter 2).

2003   MICRO-COMPUTER: Weitzmann scientist Ehud Shapiro develops the world's smallest DNA computing "machine," a composition of enzymes and DNA molecules capable of performing mathematical calculations.

2003   BREAST TUMOR IMAGING: The US FDA approves "3TP" for use in the examination of breast tumors. The brainchild of Prof. Hadassa Degani, 3TP is an advanced MRI procedure that distinguishes between benign and malignant breast growths without requiring invasive surgery.

2003   ANTI-BACTERIAL FABRICS: Prof. Aharon Gedanken becomes involved in the treatment of fabrics to prevent bacterial growth, which will eventually lead him to develop the technology for treating hospital fabrics with an anti-bacterial "coating" that will dramatically reduce hospital infection rates (see Chapter 1).

2003   CENTRINO COMPUTER CHIP: Intel Israel releases the first generation of Centrino microprocessor. Centrino is Intel's mobile computing cornerstone; it drives millions of laptop computers around the world. Successive generations of Centrino have improved laptops' function, speed, battery life and wireless communication capabilities.

2004   **TUMOR IMAGING:**   Insightec receives US FDA approval for the ExAblate® 2000 system. The system is the first to combine MRI imaging with high intensity focused ultrasound to visualize tumors in the body, treat them thermally, and monitor a patient's post-treatment recovery in real time, and non-invasively. Thousands of patients around the world have been treated.

2005   **LAB-GROWN HUMAN TISSUE:**   Dr. Shulamit Levenberg publishes the results of her work in the development of human tissue. Working with mouse stem cells, Dr. Levenberg and her partner Robert Langer produce the first lab-generated human tissue that is not rejected by its host. Dr. Levenberg goes on to use human stem cells to create live, beating human heart tissue and the circulatory components needed to implant it in a human body.

2006   **WATER FROM THE AIR:**   Researcher Etan Bar founds EWA Technologies, Ltd. In 2008 he produces a clean, green system that "harvests" water from the humidity in the air. The technology represents a boon not only to residents of water-starved desert areas, but also to farmers and municipalities around the world. Each device has the potential to provide two average American families with their entire year's supply of water without contributing to global warming or pollution.

2006   **PARKINSON'S TREATMENT:**   The US FDA approves
AZILECT, a breakthrough treatment for Parkinson's
disease developed by Professors John Finberg and
Moussa Youdim. AZILECT dramatically slows the
progression of Parkinson's in newly-diagnosed patients,
increasing the longevity of body and brain function and
improving quality of life for millions worldwide.

2007   **BEE PRESERVATION:**   Rehovot-based Beeologics, LLC,
is formed. The company is dedicated to the preservation
of honeybees — currently under threat from Colony
Collapse Disorder and vital to the world's food supply
(see Chapter 15).

2007   **AIRPORT SAFETY:**   Boston's Logan International
Airport begins testing a new runway debris detector
developed by XSight Systems. XSight uses video and
radar monitors to identify and track runway debris,
which has been identified as the cause of several airline
accidents, including the 2000 crash of a Concorde jet
that killed 113 people. XSight has the potential to save
upwards of $14 billion per year, and an untold number of
lives.

2007   **TRAUMA VICTIM STABILIZER:**   Dr. Omri Lubovsky and
his sister, mechanical engineer Michal Peleg-Lubovsky,
introduce the LuboCollar, a device designed to stabilize
trauma victims while maintaining an open airway. The
device replaces the standard procedure of intubating
trauma patients before transport, saving an average of
five critical minutes between the field and the hospital
(see Chapter 21).

2008   GIANT SOLAR ENERGY PARTNERSHIP: Brightsource
Energy, Inc., begins formalizing agreements with
California power companies to develop the world's two
largest solar energy projects.

2008   SEPSIS MONITOR: Tel Aviv's Cheetah Medical
introduces the NICOM, a bedside hospital monitor that
can detect and determine the treatment for sepsis,
which occurs in roughly 1 in 1000 US hospital patients
annually. Previously, sepsis had been treatable only after
an invasive exploratory treatment, which itself could
result in sepsis. The device is immediately scooped up by
hundreds of hospitals around the world.

2008   ADVANCED FISH FARM: GFA Advanced Systems Ltd
launches Grow Fish Anywhere, a sustainable, enclosed
and self-contained fish farming system that is not
dependent on a water source and creates no polluting
discharge (see Chapter 9).

2008   A TWIST ON SOLAR ENERGY: Yossi Fisher co-founds
Solaris Synergy, a company that creates solar energy
panel arrays that float on water (see Chapter 11).

2008   TOUGH POTATO: Hebrew University Professor David
Levy caps 30 years of research with the development
of a powerful strain of potato that can be grown in
high heat and irrigated with salt water. He shares
his findings — and discussions of where they might
lead — with scientists from Egypt, Lebanon, Jordan and
Morocco.

2009 **LUGGAGE LOCATOR:** Yossi Naftali founds Naftali, Inc. and begins distributing the Easy-To-Pick Luggage Locator, a remote luggage tag that alerts travelers when their luggage has arrived at baggage claim (see Chapter 14).

2009 **ARTIFICIAL HAND:** Professor Yosi Shacham-Diamand and a team of Tel Aviv University researchers succeed in wiring a European-designed artificial hand to the arm of a human amputee. In addition to conducting complicated activities including handwriting, the human subject reports being able to feel his fingers. Achieving sensation represents the culmination of Prof. Shacham-Diamand's work and a breakthrough in the evolution of artificial limbs.

2010 **WATER PURIFICATION:** Greeneng Solutions launches the first of its ozone-based water purification systems. Designed for commercial, industrial and domestic applications, Greeneng's product line uses ozone-infused water to eliminate germs on kitchen equipment, household surfaces, swimming pools and more. Purifying with ozone is faster and more effective than the global-standard tap-water additive chlorine, and ozone produces none of the harmful side effects of chlorine, such as asthma and contaminated runoff.

2010 **VISION LOSS TREATMENT:** VisionCare Opthalmic Technologies debuts the CentraSight device, a telescopic implant that addresses Age-related Macular Degeneration. CentraSight is the first and only treatment for AMD, a retinal condition that is the most common cause of blindness among "first-world" seniors (see Chapter 19).

2011   MINIATURE VIDEO CAMERA: Medigus, Ltd. develops the world's smallest video camera. It measures 0.99mm. The device provides for new diagnoses and treatments of several gastrointestinal disorders.

2011   HELPING PARAPLEGICS WALK: The US FDA approves clinical use of ReWalk, a bionic exoskeleton developed by Argo Technologies that allows paraplegics to stand, walk and climb stairs.

2011   BREAST TUMOR TREATMENT: IceCure Medical launches the IceSense 3, a device that destroys benign breast tumors by infusing them with ice. The procedure is quick, painless, affordable, and is conducted on an outpatient basis. Soon after, clinical trials begin to study the efficacy of the treatment on malignant breast tumors (see Chapter 17).

2011   MISSILE DEFENSE: Iron Dome, a short-range missile defense system developed by Rafael Advanced Defense Systems, shoots down a Grad rocket fired at Israel from Gaza. This is the first time that a short-range missile has ever been intercepted, opening up new possibilities for military, civil and border defense in the world's conflict zones.

2011   ENDANGERED SPECIES STEM CELLS: Israeli scientist Inbar Friedrich Ben-Nun leads a team of researchers in producing the first stem cells from endangered rhinos and primates in captivity. The procedure holds the potential to improve the health of dwindling members of numerous endangered species, as well as staving off extinction altogether.

2012   **DIABETES TREATMENT:** DiaPep277, a vaccine based on the work of Prof. Irun Cohen, is shown to significantly improve the condition of Type 1 (Juvenile) Diabetes in newly-diagnosed patients.

2012   **HELPING THE BLIND TO "SEE" SOUNDS:** Dr. Amir Amedi and his team at Hebrew University demonstrate that sounds created by a Sensory Substitution Device (SSD) activate the visual cortex in the brains of congenitally blind people. MRIs of blind people using the device show that the device causes the same brain responses that sighted people use. This discovery allows the team to adapt the SSD to allow blind individuals to "see" their surroundings by learning to interpret audio signals visually.

2012   **FUTURISTIC FOOD PACKAGING:** Israeli computer engineer Daphna Nissenbaum creates a revolutionary, 100% biodegradable food packaging material. Her company, Tiva, produces materials for drink pouches, snack bars, yogurt and other foods — all of which provide a minimum of six months of shelf life, will completely decompose in a landfill, and can be composted industrially and domestically (see Chapter 18).

2012   **THE "GOD PARTICLE":** Switzerland's Large Hadron Collider produces the Holy Grail of physics — the Higgs Boson, or "God Particle," a subatomic particle that accounts for the existence of matter and diversity in the universe. Teams of scientists from both the Weizmann and Technion Institutes were instrumental in the discovery.

Israeli Inventions
Where Were They Developed?

1. UNIVERSITIES

The Weizmann Institute of Science   Rehovot, Israel

Weizmann scientists are dedicated to solving humanity's greatest challenges. The renowned multidisciplinary research institution is differentiated by its offering of only graduate and post-graduate studies in the sciences.

1950s	Super Cuke
1954	Cancer Screener
1955	Early Computer
1956	Amniocentesis
1963	Lab-Bred Blood Cells
1966	Color Hologram
1970	Advanced Cellular Research
1972	Blood Detoxification
1977	RSA Encryption Method
1981	Leukemia Treatment
1986	Computer Language
1991	Immunology Advancement
1996	Multiple Sclerosis Treatment
2003	Breast Tumor Imaging
2003	Micro-Computer
2012	Diabetes Treatment
2012	The "God Particle"

Technion: Israel Institute of Technology   Technion City, Haifa, Israel

The oldest university in Israel, The Technion is a world-leading science and technology research university committed to creating knowledge and developing leadership to advance the State of Israel and all humanity.

1977	Digital Age Information Sharing
1981	Understanding Cellular Activity
1982	A New Form of Matter
1998	Nanowire
2005	Lab-Grown Human Tissue
2006	Parkinson's Treatment
2012	The "God Particle"

The Hebrew University   Jerusalem, Israel

The Hebrew University is ranked among the 100 leading universities in the world. It has students from 65 countries as well as 23,000 students from Israel. Founded in 1918, it is also a premier research institution.

2008	Tough Potato
2012	Helping the Blind to "See Sounds"

Tel- Aviv University   Tel Aviv, Israel

Tel Aviv University is Israel's most comprehensive institution of higher learning. Its multidisciplinary programs encourage a culture of innovation and entrepreneurship.

2007	Trauma Victim Stabilizer
2009	Artificial Hand

Ben-Gurion University   Beer Sheva, Israel

Ben-Gurion University of the Negev is one of the world's best inter-disciplinary research universities.

1967   Desalination

Bar-Ilan University: Institute of Nanotechnology and Advanced Materials   Ramat Gan, Israel

Bar-Ilan is one of Israel's fastest-growing universities. The BINA Nano-Materials Center is developing innovative methods for synthesis of size, shape, structure and chemical composition of materials.

2003   Anti-Bacterial Fabrics

2. COMPANIES

1965	Drip Irrigation	Netafim
1974	Computer Processors	Intel
1991	Baby Monitor	Hisense Ltd.
1993	Office Printer	Indigo Inc.
1993	Computer Security	Check Point
1996	Instant Messaging	Mirabilis
1997	Computer Dictionary	Babylon
1997	"Portable" Sleep Lab	Itamar Medical Ltd.
1998	Pillcam	Given Imaging
1998	First Aid	Har Hotzvim
1999	Mobileye Car Safety Systems	Mobileye N.V.
2000	Flash Drive	M-Systems
2001	Advanced Underwater Breathing Technology	Like-A-Fish
2001	Spinal Surgery Robots	Mazor Robotics

2002	Urban Air Combat/Rescue	Urban Aeronautics
2002	Terrorist Detector	WeCU
2003	Centrino Computer Chip	Intel
2004	Tumor Imaging	Insightec
2006	Water From The Air	EWA Technologies Ltd.
2007	Bee Preservation	Beeologics
2007	Airport Safety	XSight
2008	Advanced Fish Farm	Grow Fish Anywhere (GFA) Advanced Systems
2008	Giant Solar Energy Partnership	Brightsource Energy Inc.
2008	Sepsis Monitor	Cheetah Medical
2008	A Twist on Solar Energy	Solaris Synergy
2009	Luggage Locator	Naftali Inc.
2010	Water Purification	Greeneng Solutions
2010	Vision Loss Treatment	VisionCare Opthalmic Technologies
2011	Miniature Video Camera	Medigus Ltd.
2011	Helping Paraplegics Walk	Argo Technologies
2011	Breast Tumor Treatment	IceCure Medical
2012	Futuristic Food Packaging	TIPA

3. OTHER

1940s	Rummikub	Ephraim Hertzano
1948	Uzi Machine Gun	Major Uzi Gaf
1973	Drone aircraft	Israeli Army
1980s	Farm-Scale Food Storage	Agricultural Research Organization
2011	Missile Defense	Rafael Advanced Defense Systems
2011	Endangered Species Stem Cells	Scripps Research Institute